George Spe[...]

animals at war

Carolyn Barber

animals at war

Harper & Row, Publishers
New York, Evanston,
San Francisco, London

Animal Life Series

Series editor: **Cathy Jarman**
Zoological consultant for this volume:
Robert Burton

The chapter on "Animals Enlisted by Man" was contributed by Richard Humble

ANIMALS AT WAR. Copyright © 1971 by Macdonald. All rights reserved. Printed in Great Britain for Harper & Row, Publishers, Inc. No part of this book may be used or reproduced in any manner whatsoever without written permission except in the case of brief quotations embodied in critical articles and reviews. For information address Harper & Row, Publishers, Inc., 49 East 33rd Street, New York, N.Y. 10016.

FIRST U.S. EDITION
STANDARD BOOK NUMBER: 06-010208-X (cloth)
STANDARD BOOK NUMBER: 06-010209-8 (paper)
LIBRARY OF CONGRESS CATALOG CARD NUMBER: 76-185620

Contents

1	The games of war	7
2	Trials of strength	27
3	The rules and regulations	53
4	The hunter and the hunted	85
5	Animals enlisted by man	109
6	Man – what went wrong?	127
	Indexes	149

Editor's note

The title of this book, 'Animals at War', perhaps suggests that animals actually go to war and wage bloody, destructive battles to kill their fellow beings.

There is, indeed, fighting in the animal kingdom but this is to protect family, home or mate and animals rarely kill except for food. If animals used their weapons of teeth, claws or horns for fighting and killing their own kind, a whole species might destroy themselves. Instead, animals have 'rules' which prevent serious injuries. The methods they use are fascinating, ranging from the head-on clashes of musk oxen to the shrieking battles between shrews, and a victor emerges without harming his rival.

Man is different – he has little restraint in battle and his wars are full of needless killing. He has even used animals, from horse to carrier pigeon, to help give him victory.

This and man's attitude to war, with the differences and similarities to the ways animals fight, are fully discussed in Carolyn Barber's beautifully illustrated and informative book.

To present the other side of the picture Alison Maddock has written a companion book, 'Animals at Peace', which deals with the peaceful aspects, the family life, and helpful associations which are quiet ways of life in the animal world.

The games of war

A man who does not love nature is a disappointment to me, I am liable to mistrust him.

Otto von Bismarck

It is impossible to compare the destructive nature of man's warfare with the war found in the animal world. To animals war means a struggle for survival: competition with rivals and pitting wits against predators is just part of their war. They must also adapt to the changes in surroundings imposed by man.

Animals do not go to war in the way that man does. They do not march into battle in armies with the aim of defeating an enemy. But many aspects of their lives are concerned with war of a kind: the harsh, competitive struggle to survive.

Not that the life of animals is one unending round of bloodshed and violence – the idea that nature is always 'red in tooth and claw' can be misleading. Survival is not achieved without a fight, but this fight has many facets, both visible and invisible. For example, the violence of a rabbit's death as it is snatched up in the eagle's talons is immediately obvious, while the means by which animal species overcome the onslaughts of climate are not so easily seen. Man's battle waged with machines and pesticides against the creatures which live off his agricultural crops is certainly like a war, but the open fighting between rival male stags during the rut is more like a duel or a trial of strength. Each struggle has meaning and purpose in nature, and each has a place in this book.

The behaviour of animals is of especial interest in this era when man is trying with increasing urgency to understand the origins of his war-like nature. Some kinds of animal strife have attracted the particular attention of students of animal behaviour. These are the clashes which occur between the members of one species, and are technically called intraspecific or within-species fighting – a kind of 'civil war'. A fight between two lions is an

Zebra civil war. The fights that break out between two animals of the same kind, between zebra and zebra, is called intraspecific *fighting. The encounter is usually over territory or a mate.*

example. There are also several kinds of fighting when members of two different species come up against each other. These are termed interspecific or between-species encounters. An example is the relationship between hunter and hunted, that is, predator and prey.

The strife within a species is an expression of the important inborn characteristic called aggression. Although in international affairs this word often implies the initiation of hostilities, in behaviour studies it has a broader meaning. An aggressive act consists of inflicting damage on another animal, or at least threatening or attempting to do so. A tendency to be aggressive is behind the outbreaks of fighting among animals of one species, but it is not usually considered to play a part in interspecific fights. Although we cannot know the emotions felt by other animals, it seems unlikely that a predator feels in any way aggressive towards its prey, but is simply obeying the impulse to obtain food.

Man's problem

At first sight aggression may seem a curious attribute to have become part of the species inheritance. In the course of evolution, the features which are selected for and become more common are those which are of advantage

When a lion attacks a warthog, or any animal fights another animal of a different species, it is known as interspecific *fighting – war between a predator and its prey.*

in ensuring the survival of the species. For the species to continue, a certain number of individuals from each generation must live to reproduce, so that any character which promotes individual survival is of value. If animals fight among themselves and even kill each other, how can the welfare of the species be served? The answer, as shall be seen later, is that aggression is of great value in the spacing out of individuals, provided injury is kept to a minimum. In animals, aggression is under strict control: fights have developed in such a way as to give the least risk of damage to each opponent. In man, however, control seems to have been lost, and aggression has got out of hand. By studying animals, behaviourists are hoping to understand and even cure the runaway destructiveness of the human species.

War among men is only too apparent in this present century. Two terrible World Wars have been fought and,

today, fighting breaks out in many countries from Israel to Ireland. Man's unbridled aggressiveness is not only obvious during warfare; it is with us all the time – in football crowds, street fights, even bus queues and traffic jams. But such strife is set apart from the rest of the animal world, because it is an almost unique example of members of a species banding together to fight and kill others of their own kind. We can be sure that such behaviour is not instinctive; biological evolution could not have produced a built-in drive to destroy the species.

In trying to understand how this situation can have arisen, we must remember that man, although part of nature and subject to the same biological evolution as other animals, is separated from other living creatures by his culture. The beliefs and customs which he learns are an important way of adapting to the surroundings and have exerted a powerful influence over his development.

The end of the match and Celtic have defeated Rangers by two goals to one. It is customary for the victorious side to do a lap of honour, holding the Cup aloft. But as Celtic parade the League Cup thousands of Rangers fans invade the pitch heading straight for the Celtic players. Scuffles, blows, yells and the inevitable arrests follow. Such a scene as this has become commonplace today and is just another example of man's uncontrolled and uninhibited violence directed at his own kind.

Cultural evolution proceeds at a much faster rate than biological and sometimes comes into conflict with it. Warfare developed when groups of men became separated from each other by their cultural background, perhaps even ceased to speak the same language, and began to compete with each other for economic and other reasons. So, while the study of animal behaviour is invaluable to an understanding of man, explanations of man's actions cannot be inferred in a simple fashion from animals to man. His are more complex problems, as the last chapter of this book will show.

The car is one of man's territories. In traffic jams when tempers flare up very easily, each impatient driver feels safe enough within the confines of his car-territory, to swear or shake his fists at the neighbouring driver.

Displays of threat

In a survey of the open fighting which takes place in the animal world, the most striking fact which emerges is that there is very little large-scale antagonism. Intraspecific fighting most often happens in the breeding season, when the males are particularly aggressive and are looking for mates and a place to set up home for the rearing of the young. Often quite spectacular, the tussles of rival males attract attention to the unrest among animals at this time. Lions, hippopotamuses and elephant seals are among those which will fight with some violence, but deaths are rare. Animals do not fight each other unless it is absolutely necessary and then never with the intention of causing death. Bloodshed is usually accidental.

In fact, the species which engage in open combat are in

the minority. Surprisingly often, the contestants do not actually come to blows. The controlled aggression is seen in the form of a convincing display of threat, so that the so-called fighting consists of sequences of postures and expressions recognised by each opponent. The displays are sometimes very elaborate in birds such as gulls and in some fishes. When the fight is over territory – an animal's home ground – the outcome can often be predicted; the territorial owner will most likely be the victor, the intruder the loser. There will be more about these territorial disputes in later chapters.

Not all intraspecific fighting is directly concerned with competition for mates. Many animals which live in groups will fight and threaten each other to achieve a position of leadership. There may actually be an order of rank, or 'peck-order', within the group which reflects the relative success of each animal in disputes. The top, or dominant, animal submits to no-one; he is regularly the victor. The next in rank dominates over all the animals except the leader, and so on down the peck-order to the lowest animal, which is defeated by all the other animals and defeats none of them. This kind of social order is found in a number of animal species. It is usually referred to as a peck-order because it was first observed in domestic hens which fight by pecking and nipping at each other. For a hen to change her position within the peck-order she must win a fight with a superior hen. It is easier for female jackdaws to alter their rank within their social order. Should a female of low rank become engaged to a male high up in the scale she immediately rises to the same rank as her betrothed. Monkeys too have a strict social order. In a troop of langurs there will always be one animal that leads. The amount of actual fighting after the initial phase of establishing the peck-order is reduced by displays of threat. Systems of rank in fact, help to regulate the life of the group and to keep it together. Once again, aggression is channelled and put to constructive use.

After some throat swelling preliminaries male goannas battle with tooth, claw and tail. The fight may look fierce but victory and defeat will be decided without death – bloodshed rarely occurs.

Squabbles and quarrels

The causes of many fights are not immediately obvious. Giraffe live in small herds with no definite territorial behaviour, and the males may fight even when there are no females in the vicinity. Sweeping their great necks, they slam their heads at each other's flanks, using tremendous force. Their blunt horns do little damage; their use is in fact an evolutionary mystery. After about fifteen minutes, one of the participants gives up and moves away. Observers are not certain of the reasons behind this duelling, but the giraffe may well be establishing an order of rank for the area.

Squabbles for food are often seen between animals. Gulls or ducks fighting over bread are a common example. Disputes are of short duration and among animals with a social hierarchy they are settled by rank. The more dominant of two baboons, for instance, will have first choice at any new food source.

Certain species of animal are sometimes described as being more quarrelsome than others. Both male and female shrews will fight each other whenever they meet, except for the purposes of mating. This probably has to do with a shrew's voracious appetite. These little animals must feed almost continuously in order to stay alive, and so all potential competitors in a particular shrew's feeding area must be driven out. Unjustly perhaps, the shrew's reputation has led to the term 'shrew' being applied to any spiteful, troublesome woman, as in William Shakespeare's *The Taming of the Shrew*.

It is very rare for animals to become cannibals or murderers. Occasionally a disturbed mother mammal may kill her offspring. One of the few natural examples of killing occurs in colonies of honeybees. In the autumn the drones are turned out of the hive to die, while the rest of the colony remains to hibernate. This is not a suicidal move for the species, however, because the drones are males which do not work and have no further use after they have fertilised the eggs of the queen bee. Similarly, when a female mantis eats her mate he has already served his purpose and the action, though bizarre and distasteful to us, does not endanger the continuance of the species.

Giraffes settle a matter of dominance by 'necking'. Their short, blunt horns do little damage in fighting.

Seagull fracas. These quarrelsome birds are always ready to fight over food thrown from boats.

Overcrowding – sika deer study

Killing does occur when conditions in an animal population have become abnormal, especially when there is overcrowding. This may happen among captive animals. Competition for food becomes very intense and aggression more and more marked. One overcrowded group of hippopotamuses was studied and found to have a very high incidence of fighting. In a way this is adaptive, because deaths will allow the population to return to its normal, economically viable, size. Frequently, however, the cause of death in overcrowded populations is not fighting but stress. This seems to have a lethal effect on the adrenal and other glands. The loser in a territorial dispute between caged rats is unable to escape, and though it may appear to be uninjured it later dies – presumably from stress.

Although sika deer have antlers with which they could battle, they often rise up on their hindlegs and fight with their forelegs. Sika deer were the subject of an experiment on overcrowding and how animals solve this problem. The experiment took place on James Island in Chesapeake Bay, U.S.A. where four or five deer were released in 1916. John Christian went to study the deer in 1956 by which time the herd was nearly three hundred strong.

In 1916, four or five sika deer were released on James Island in Chesapeake Bay, United States. When the scientist John Christian went to study the deer in 1956 the herd had grown to nearly three hundred strong. He believed that animal populations are regulated by mechanisms which respond to the density of animals. In the first three months of 1958 over half the deer died. By 1959 the decrease had continued until the population had levelled off at about eighty. The question was: why had this rapid fall in numbers occurred? Although the deer were crowded there was still adequate food, and a sample of animals which Christian examined appeared to be in good health, with shiny coats and well-developed muscles. However, a comparison of the internal organs of animals which had died in the population crash was made with those of animals examined when Christian first arrived on the island. The important finding was that the adrenal glands, which are near the kidneys and ensure an effective response to stress, were forty-six per cent heavier in the deer which had died in the crash. This was a strong indication that they had died from stress as a result of a natural population-control mechanism.

Animals will often take to quite drastic methods to regulate their population. Flour beetles, when overcrowded, will produce a gas which is both lethal to their larvae and inimical to mating. Fishes, crabs, lions and many other animals kill and even eat their young when conditions become intolerable. Some species avert threatened starvation due to overcrowding by setting off on long migrations. One of the most famous is the mass movement of lemmings, which are stopped by no obstacles in their frenzied rush, not even the sea, where many of them may drown.

For man's pleasure

Man's over-aggressive nature may be due in part to the effects of crowding. Another useful comparison may be made with certain animals which do fight very viciously, and often to the death. These are fighting cocks and the famous Siamese fighting fish. Males of both these species fight in the wild, but their aggression has been enhanced by special breeding so that it is unnaturally strong. In parts of the world where the fighting of these animals is encouraged for purposes of sport, the most aggressive animals are deliberately selected and bred for fighting ability. Perhaps man is the product – or the victim – of his own 'artificial selection', which decrees that an aggressive personality is necessary for survival.

In the wild, Siamese fighting fish – which occur in

rivers, lakes and ditches all over southeast Asia – rarely keep up their fights for more than fifteen minutes. But the specially bred varieties which are used in sport in Thailand are considered poor specimens if they fight for less than an hour, and some will battle on for up to six hours. The contest between two of these fighting fish can be quite dramatic to watch. When two males are put together in an aquarium their colours heighten and they take up their positions, swimming side by side with one fish slightly in front of the other. With fins erected and gill-covers expanded they attack with lightning speed. They try to bite off each other's fins and patches of scales from the flanks. Sometimes they will meet head on with jaws interlocking. This deliberate injuring in battle shows how the normal behaviour has been distorted by breeding for fighting prowess.

A common tourist attraction in India and Egypt are the staged fights between the cobra and the mongoose. These animals would not normally fight each other in the wild as neither preys upon the other, but when put together within a small enclosure, surrounded by onlookers, their first reaction is to fight. Left alone the battle could go on for quite a time but after about fifty minutes the

Bred especially for the sport they provide, two vicious and strong fighting cocks battle to the death.

Opposite. *Rearing and snorting a pair of chestnut horses sort out a quarrel over dominance. Their backward-facing ears and curled lips indicate aggressive feelings.*
Overleaf. *Leopards like to store their surplus food high in a tree, usually positioning it in a convenient fork well out of the way of scavengers such as jackals.*

18

animals are separated to prevent their deaths and keep them for the next tourists.

Attacker and attacked

War between animals of different kinds is usually a simple matter of getting enough to eat. It is not concerned with special postures and displays, but with the physical ability of each opponent, and the weaker animal usually falls prey to the stronger. The hunter pursues the quarry until it is cornered or can run no more. Then the hunted creature may well give in without a struggle, but sometimes it will fight to try and save its life. There is no anger involved in this type of war; it is a life-preserving function

A common tourist attraction in India and Egypt is the staged fight between a cobra and a mongoose. The fight is usually very one-sided with the advantage towards the agile mongoose. When fighting the mongoose erects its hair so that it looks twice its natural size. This makes the snake strike short. The cobra raises about one third of its body above the mongoose. But its strike is slow so the lively mongoose has a chance to sidestep the snake and bite the snake in its neck. The fight is stopped after about fifty minutes to prevent the two animals from killing each other and to keep them alive for future tourists.

whereby one animal is food for another. A lion about to kill a zebra does not look like one threatening another lion. In the latter event, there is a menacing attitude with growls and laid-back ears which is quite absent from the hunter–hunted situation.

Another example of interspecific war occurs when an animal that is closely guarding its home encounters a potential enemy, like the interaction between an eider-duck and a herring gull. The gull is really in the position of a predator threatening to make the duck's young its next meal. Whenever a gull comes too near, the ducklings quickly gather around their mother. The eider-duck puts up a fierce and vigorous defence against the attacker. She stretches up her neck and raises her wings ready for action. The herring gull is soon discouraged, for as long as the ducklings remain in a close circle around their mother they are safe. It is when they scatter that the herring gull can easily kill one for a meal. Defence and attack of this sort are the subject of a later chapter.

Mobbing is another effective way of warding off potential enemies. Jackdaws will mob an owl if it appears in the daytime, to persuade the owl to find a hunting ground elsewhere. Small birds will band together to mob birds of prey and weasels. Animals which hunt singly have a chance of catching and killing their next meal only if they take their prey by surprise. A fox followed through a wood by a loudly screaming jay will have little chance of hunting successfully in that wood for some time.

All forms of animal fighting, whether intraspecific or interspecific, serve in the long run to keep the species going from one generation to the next. This may appear contradictory at first. How can interspecific strife, which often ends in the death of the prey, be of advantage to the prey species? The hunter–hunted relationship is a well-balanced one; no animal is preyed upon to extinction, as this would be disadvantageous to both predator and prey. A tiger, for instance, kills a gazelle or other antelope only when hungry – not for the sake of killing. If the tiger did kill for pleasure it would not be long before all the gazelles were destroyed, and the tiger would starve to death. The tiger would bring about its own destruction in the end.

An eider-duck defends her brood against an attacking herring gull. The mother stretches up her neck and raises her wings in an attempt to save her young. Providing the ducklings stay close to their mother they are safe. It is when they scatter or are not able to reach her in time that the predatory herring gull will swoop down and take one for its next meal.

On the other hand, if there were no factors such as predation to limit the gazelle population, numbers would rise to uneconomical proportions, leading to overgrazing and even starvation.

Marching to war

It is difficult to find analogies with our well-equipped, well-disciplined armies in the animal world. There are a few species which do occasionally move in large numbers. Some social animals, such as wolves, co-operate by hunting in groups. Locusts, serious pests in many parts of Africa and Australia, form huge swarms in some years which devour everything green in their path, but they do not destroy each other. Migrating lemmings, hastening onward in their agitated search for more space, do not turn against each other. Certain nomadic ants provide what is probably the closest animal equivalent to the armies of men.

The army or legionary ants of tropical America, and the driver ants of Africa, have the reputation of being exceptionally savage predators. Progressing in columns or as seething masses on a wider front, they march in search of food, tearing to pieces almost all the living creatures they come across – including other species of ants. Even young birds in their nests are not safe, and tethered or sleeping animals and men may be quickly reduced to skeletons. Army ants have regular phases of marching activity, related to the queen ant's reproductive cycle and to the stage of development of the larvae, which are carried by worker ants when the colony is on the move. There is no fixed nest; fresh bivouacs are made each night from the clustered bodies of up to a hundred thousand workers, and contain chambers and passages for the queen and the larvae.

Like many other species of ants and termites, the army ant colony contains certain large forms with big jaws, known as soldiers. The soldiers are specialised worker ants. They seem deliberately to take up a position at the front and sides of the marching column, but this is probably simply a result of their larger size making it difficult to walk among the other workers. The army ants also seem to mirror human tactics as they divide into wings to outflank their prey, but even this seems to be an accidental result of following the scent trail laid down by the leading ants, and does not indicate any deliberate organisation.

Ant behaviour is in many instances quite alienated from that of most other animal species. In the ant world are found slave-making species, which raid the nests of other kinds of ants and take the eggs and larvae, to be reared as workers in the raiders' colony. More significantly, there are also colonies which wage civil war on their own species, attacking and fighting often with terrible carnage. Here is a real parallel with the warfare of mankind. Why it should have evolved is not certain. Ants are very numerous, and the loss of numbers of the non-reproductive workers must be offset by advantages to the species in competitive warfare. A pioneer of ant behaviour studies wrote that 'the ant's most dangerous enemies are other ants, just as man's most dangerous enemies are other men'.

But the ant is only one of thousands of animal kinds. It is quite impossible to imagine bears or elephants forming well-ordered regiments; or, for that matter, birds, lizards or frogs marching in squadrons towards an enemy with the sole purpose of killing. Apart from the ants, infinitely far removed from the human species in the regulation and organisation of their life, that sort of activity is practised only by man.

Animals versus environment

Quarrels and clashes are not the only form of war that animals experience in their lifetime; they are constantly engaged in unseen struggles with their surroundings. To

survive changes in climate, to live in extreme heat or cold, to find food and water, to resist diseases and parasitism; these are some of the obstacles animals are up against. Various physiological mechanisms keep the body's internal state constant in the face of external changes. Among them are the water-balance mechanism and the temperature-regulating system of warm-blooded animals.

Animal species that live in desert regions must avoid drying up and must at the same time keep cool. They must adapt to drought and heat. When water is in short supply an animal's body must resolve by physiological means the conflicting needs to conserve water for vital functions and lose water to keep cool. The smaller animals avoid climatic extremes by seeking shelter during the day. Animals that inhabit extremely cold areas, up mountains and in polar regions, have a cold and wet environment to contend with. Nevertheless, the animals are well adapted to their way of life and have overcome the difficulties of living in what may seem to many of us to be inhospitable places. In winter, animals of the cold lands may survive because they possess a thick coat of fur, or they may migrate to warmer climes, or hibernate until the spring.

Men and women, too, have managed to survive and live in both extreme environments. Man is fundamentally a tropical species, but he can adapt to all climates by means of clothing and housing, and by physiological means – sometimes acquired through acclimatisation and

Animals that eat crops or spread diseases among crops are labelled pests. Pests come into constant conflict with man who uses chemical weapons, such as pesticides, to fight them with.

sometimes built-in to the genetic make-up through the process of evolution.

Many animal species, especially insects, have been subjected to human chemical and mechanical warfare because their interests clash with those of man, and most of them have evolved adaptations accordingly. All kinds of creatures have been labelled pests because they attack crops, spread disease, or parasitise domestic animals or man himself. Mankind cannot be blamed for wanting to produce more food and rid himself of pestilence, but during this century it has become clear that the means employed have at times been ill-advised. Indiscriminate destruction of habitats and the use of pesticides have created a serious pollution hazard which threatens both man and animals.

One of the species singled out for extensive attack is the mosquito, carrier of several diseases including, in Africa, sleeping sickness. It is now clear that D.D.T., one of the chemicals used against the mosquito – and also against a variety of other pests – can find its way through food into the human body in dangerous amounts. In addition, insects sometimes become resistant to its effects.

Another important case of resistance concerns rats and the chemical called warfarin used to control them. Warfarin was first used as a poison against rats in 1952. In 1958 it became apparent that there were populations of this rodent that were not being killed by the chemical. Today there are such populations in England, Scotland, Wales, Holland and Denmark. It is only the common rat that does not succumb to the poison; warfarin is still lethal to the ship rat. The rat problem is confined to certain localities and does not affect the population as a whole, but the resistance of house mice is a more widespread problem. Warfarin kills slowly. It prevents the blood from clotting and the affected rat dies from an internal haemorrhage. Other stronger poisons could be used but they do not have the same advantages as warfarin. With a stronger poison great care must be taken that other animals do not eat it or eat a rat that has died from it.

A rat does not become resistant to a chemical overnight. The resistance may exist in rat populations for many generations but is not revealed until a poison such as warfarin comes into regular use. Then only the animals most able to survive the toxic effects live to breed. These breeding rats will produce more warfarin-resistant offspring than non-resistant, so eventually a population will consist entirely of resistant animals.

So the animals fight back: species which have not the make-up to withstand new conditions go under. But the impact of modern man's activities on wildlife is immense, and may prove too much for many quite harmless kinds. The survival of animal species all over the world is now threatened on a wider scale than ever before.

Trials of strength

...the parent birds seem to maintain a jealous superiority, and to oblige the young to seek for new abodes: and the rivalry of the males, in many kinds, prevents their crowding the one on the other.

from The Natural History of Selborne by Gilbert White, 1778

An animal's home is his castle and must be guarded against intruders. The sparring matches that so often occur between animals in the breeding season are usually for the defence of territory. Musk-oxen battle with their heads, tree shrews shriek at each other and gourami fishes mouth-wrestle.

Love and war are recurring themes in the history of mankind. In a sense, these intermingled themes are also very important in the animal world. Aggression within a species reaches a peak in the breeding season, for it is then that the males challenge each other. Sparring matches break out as rivals clash over the possession of females or a particular territory. Sometimes the females fight among themselves as well.

This fighting between members of the same sex is called reproductive or rival fighting. Such contests between men rarely happen today. Romantic legends tell of gallant knights duelling for the hand of a fair lady. But as the modern woman plays a more prominent part in society many men no longer feel she is a damsel in distress to be won by strength and courage. In animals, however, rival fighting is clearly an important and valuable aspect of behaviour, and occurs in many species with seasonal regularity.

For animals to breed successfully, to rear young so that they can eventually fend for themselves, a home base of some kind is often necessary, especially among vertebrates (animals with backbones). A home may represent stability and security, a place where food is available and easily acquired. This also applies to human beings. Most men and women, whether they lead a city life or tribal existence, enjoy the comfort of their own home and regard it as a place of love and friendship, where there are familiar surroundings made individual by personal possessions. A home for an animal is often situated in a defended piece of land or water known as its territory. Members of the same species are jealously excluded from the area. Animals that at any other time of the year would tolerate the presence of fellow members of their species – for example shoals of fish, flocks of birds and herds of antelope – seek isolation from their companions during the breeding season.

An animal's territory may cover an area of miles, as the golden eagle's does, or it may be only a few feet in diameter, like the herring gull's. It often centres on a nest, burrow, den or other home. However large or small this territory is, the owner will defend it vigorously against any intruder. Established in the first place by the male at the beginning of the breeding season, the territory will later be defended by the female too if the species is one which pairs for any length of time.

In species in which one male has several mates, the fighting for possession of a harem or group of females is closely related to territorial disputes. The area containing the females is the territory which the male defends. The fighting between male rivals never purposely ends in death; there is no point in this. The rightful owner of the territory or harem of females merely wants to show the intruder off the premises and employs various threatening postures and expressions to do this. These actions are

Some fighting fish beat their tails sideways without touching, the water-movement creating a stimulus like a physical blow.

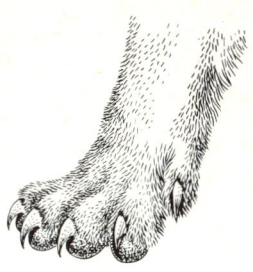
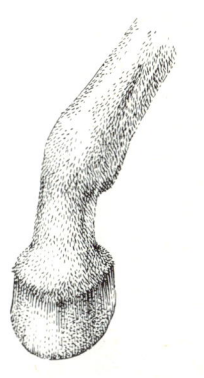

easily and quickly recognised by his opponent, and if the opponent is wise he withdraws without much ado. Sometimes the dispute does erupt into a fight, but even when there is apparently violent physical contact severe damage is rarely inflicted. Fatalities do happen from time to time, however, and this is exactly why the threat of injury remains effective even if seldom carried out.

Animal weapons

As each kind of animal fights in its own individual way, so the weapons that it uses are suited to its method of combat. Dogs bite, gulls peck, horses kick with their forelegs, deer push with their antlers, elephants fight with their trunks, and crabs battle with their claws. Some fish bite – the salmon has a jaw well suited to this – but most fish fight by sending strong jets of water at their opponent by vigorously beating their own tail from side to side. These water movements stimulate the sensory organs along the flank of the offending fish, and so the message is received without either fish making contact.

Teeth, claws and horns – these are the most effective mammalian weapons. The original functions of teeth were to seize food and reduce it to a manageable size. Claws protected the ends of the fingers and toes. Just by an increase in size, shape and sharpness teeth and claws are easily adapted to use as weapons in combat. Mammals such as the rhinoceros, deer, antelope, sheep and goat do not have teeth and claws as their weapons, for in these

Mammalian weaponry: a cat's claws, a horse's hoof and a wolf's fierce teeth can all be used as weapons in a fight. Like many of the hoofed mammals, rhinos are equipped with horns for this purpose.

animals the biting teeth are usually reduced and they have no effective claws. They are not left helpless, however, for the males (and sometimes the females) have horns or antlers for their defence and attack and sometimes they give hefty blows with their hoofs. Male sika deer will rise up on their hindlegs and battle with their forelegs, using their hoofs as weapons. The rhinoceros, though, has feet that are protected by hoof-like nails, adapted for carrying a great weight and not for use as weapons but, possibly to compensate, it has one or two large nasal horns which are effective battering rams. In fact these horns seem to be as important in courtship activities as they are in fighting. The three Asiatic species of rhinoceros have only small horns but their incisor teeth are developed as tusks. This modification occurs in some deer as well. The musk deer, the water deer and the chevrotain have their upper canines adapted for use as tusks that project between the lips when the mouth is shut.

Among the antlers of antelopes and deer, which are often used for defence and attack, there is a great variety of shape and size. In most deer species the rival fight – an unusually noisy and rough one – follows a set pattern. Red deer stags in Great Britain begin to get restless in September at the start of the breeding season, a time known as the rut. They find their voice in a belching roar known as belling, with which they advertise their rights over a group of hinds. Hindless stags may challenge the owner, but only males with equal strength antlers engage in

Left. *An oddity of the deer world is the tiny musk deer, whose upper canine teeth take the form of long tusks, perhaps in compensation for the lack of antlers which are the characteristic weapons of most other deer.*

Right. *The boisterous pre-mating fights of hares have given them the name of 'mad March hares'. Boxing, kicking with the hindlegs and jumping, the animals sometimes inflict fatal injuries upon each other.*

battle. Animals with small or weak antlers will not fight a stronger deer but acknowledge inferiority. The males do not charge furiously at each other but feel cautiously with their antlers, and then at the last moment thrust with their heads, the forehead coming nearly parallel with the ground. Both animals are aiming to stab their opponent in the ribs, but as both are also taking avoiding action the result is a head-on shoving match. With antlers intertwined they wrestle with each other, pushing this way and that.

The fights or rutting battles of the stags are often waged with great ferocity and deaths sometimes occur. This is one of the exceptional examples of a fight to the death in intraspecific encounters. Death is not an intentional outcome of a fight between stags. It may happen that the antlers become so wedged together they are impossible to separate, in which case death through exhaustion and starvation is inevitable for both animals. This irreversible interlocking of antlers can occur only with the more highly-developed and complicated type of antler found in the red deer and in the elk of North America.

The horns of the oryx do not branch like the red deer's, but are gracefully curved. They are extremely sharp and capable of inflicting severe wounds and are very effective against enemies. In rival fighting, however, the oryx antelopes – which live in Africa – battle with their heads in such a position as to safeguard against unnecessary injuries, the horns pointing vertically upwards.

Two scarred bull elephant seals threaten with loud roars before falling upon each other with what appears to be great ferocity. The wounds inflicted by the teeth of the rivals are, however, quick to heal.

Duelling rules

The formidable musk-ox of the Arctic wastes is another mammal that battles head first with all its might. When two bulls meet during the rut, they will dash at each other as on a given signal, their heads well down, and clash with a terrific crack that may be heard over a mile away. This does not, as would be expected, knock the two animals senseless. The enormous impact is taken by the flattened bases of the horns at the front of the skull. The battle is over only when one bull surrenders, driven away to lead a solitary, wandering life.

Fighting is also very stereotyped among male elephant seals, which gather harems of females rather like deer. One male in particular is the master bull, and he will fight any male which challenges him. Roaring loudly, the contestants rear up their bodies to a height of eight or nine feet and inflate the trunk-like structure of the nose. Then they attack each other, each trying to tear at the other's body with their teeth. Gashes and serious cuts are not uncommon, but the tough males seem to recover without permanent injury.

Quite different from the charging musk-ox and the

bellowing elephant seal is the duel between two male tree shrews. These inconspicuous mammals from India and China can make a surprising amount of noise for their small size. Their contest is a battle of nerves. An intruder is immediately bombarded with shrill staccato shrieks, which is usually enough to make him retrace his steps very rapidly. If the intruder is in full fighting fettle, however, he too begins to shriek and both the shrews rear up on their hindlegs kicking each other with their forelegs. If this does not make the offending shrew go away the territorial owner flings himself on his back and squeals even louder, continuing to do so until the intruder can stand it no longer and finally runs away – victory by dogged persistence.

Often in rival fighting the more dangerous the weapons used the more exacting and scrupulous are the rules. An animal may have a weapon which is capable of inflicting mortal injuries. This is a great advantage when attacking another species or defending itself against predators, but means that special care is needed during fights with its own kind. Sometimes the weapons which are employed in intraspecific fighting are not the same as those used against predators. A giraffe, for instance, duels with its blunt horns, but defends itself with its hoofs. Rattlesnakes from North America are another good example. The bite they use against predators is just as dangerous against a rival. It is curious that many snakes are sensitive to their own poison. It would be quite easy during a rival fight for one snake to bite out viciously at its challenger, putting a quick end to the battle. But death is not the purpose of reproductive fighting and so fights follow certain rules and regulations that are closely adhered to by the two contestants.

In the case of the snakes a wrestling match begins with them gliding round each other in diminishing circles until they lie alongside each other. Then in a twisting motion the hinder parts of their bodies clasp each other, at the same time they rear their heads and front parts of their bodies to a height of about twenty inches. This seemingly affectionate movement is deceptive for, in fact, each combatant is filling its lungs with air and pressing with all its strength against the other until they suddenly break away. The snakes jerk apart like springs and beat their bodies on the ground. Each snake rises as quickly as possible and continues to press the other sideways. Usually one of the two snakes manages to exert enough pressure just behind the other's head to force it to the ground, where it holds it fast for several seconds. This is the end of the fight. The victor relaxes its hold and allows the loser to escape and make off as quickly as it can. By following this set pattern both snakes come through the battle none the worse for wear, yet a decisive result is achieved.

Wrestling rivals

Males of many species engage in reproductive fighting. From top to bottom, crickets, chaffinches, rats and Australian brown snakes.

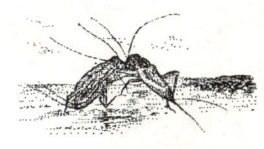

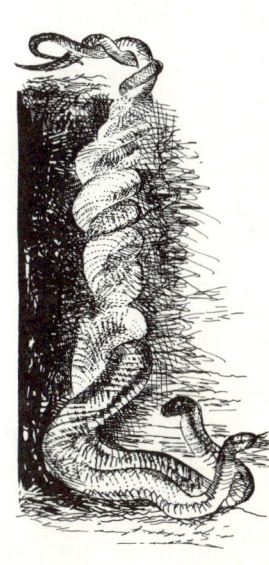

When two of these African giants come together, the contest is a formidable trial of strength. As usual, the fight is over the possession of a female.

The biter bit

The rivalry between male sand lizards of Eurasia is most unusual. A fight is preceded by the two animals posturing to each other, arching their backs, puffing out their necks and lowering their heads. After this introductory display one male grasps and bites his rival in the neck, quite vigorously but without drawing blood, and holds him fast for a while in this uncomfortable position. The secured lizard waits patiently without struggling until it is released. This lizard then takes its turn to have a bite and the contest goes on until one or other gives in. The peculiar part of this contest is that the victor is not the biter but the bitten. It is a test of the capacity of one lizard to tolerate the bite of another not a test of the strength of the bite. The defeated lizard does not actually turn and flee, but carries out a symbolic flight, turning its tail on its victor and dancing up and down on the spot for a while. This is accepted as a sign of surrender and hostilities are immediately dropped.

Birds are strongly territorial animals. Their fighting is ceremonial and closely follows a sequence of events recognised by each opponent. Although many are social animals during most of the year, in the spring they will untiringly defend their home ground against intruders.

Ruffs are sandpiper-like birds of northern Europe and

Asia with a spectacular fighting display. The males have a 'ruff' of feathers which gives them their name. When two ruffs fight they peck at each other, jumping into the air kicking one another, hitting with their wings and seizing each other with their bills. Often they come face to face with heads lowered, ruffs and head tufts raised, making downward pecking movements, or they may run wildly about turning this way and that. This agitated running around, often running parallel to the rival, is found in many birds and indicates the fear which is mixed with the birds' aggression.

Some birds show territorial behaviour all the year round, and some seem to take pleasure in fighting. The mynah is the extremely quarrelsome starling of southern Asia. Mynahs tussle on the ground, and the fight begins with a great deal of leaping about, wing flapping and low cries, followed by much kicking and pushing with the feet. Coots, too, are renowned for their aggressive quarrelling, and not only during the breeding season. Their arguments take place on the ponds and lakes where they live. The first sign of a fight is when one or both rivals take up an aggressive posture, head down and wings arched. If the rival does not retreat then they swim together, suddenly erupting into a water-throwing match, sitting on the backs of their tails and splashing water over each other by beating with their wings and feet. The fight is usually over in a few seconds and neither bird is injured. Coots are so keen to enter into combat that even if one retreats post-haste it is often pursued by the other.

The 'kissing' fishes

Just as the mynah goes in for foot-wrestling, many fishes engage in contests of mouth-wrestling. In the aquarium fish known as the kissing gourami, for instance, it is not very clear whether the ritual of 'kissing' is an aggressive action or part of courtship. It is probably a bit of both, because it is performed by a male and a female as well as by two males. By putting two male kissing gouramis in an aquarium large enough for each one to stake out a territory, it is quite easy to watch the two fishes take part in rival fighting. The fight begins with two fishes facing each other and swaying backwards and forwards. They then come together their thick lips firmly joined in an exaggerated 'kissing' action, each one testing the strength of the other. As in most territorial disputes it is the territorial owner that usually wins.

A very aggressive and once popular aquarium fish is the brightly-spotted Jack Dempsey, so named after a world heavyweight boxer. This native of South American rivers grows to about eight inches in length. Because of its quarrelsome nature this fish causes havoc when put in a tank with fish of another species.

During the breeding season the males establish a territory. If this territory is violated by another male the fight goes as follows: the territorial owner faces the intruder and begins a sideways display, swimming alongside the newcomer so that the two fishes are head to tail. The owner raises his dorsal and anal fins, spreads his paired fins and raises his gill-covers; at the same time his colours become brighter, so that viewed from the side the fish looks very

A familiar human action performed by a pair of fish. No-one is quite sure whether the kiss of kissing gouramis is an affectionate gesture or, more likely, an act of territorial aggression.

Opposite. *A magnificent red deer stag, with an impressive display of antler power, challenges rival stags in the rut by belligerent and repeated belling.*
Overleaf; top. *Green lizards use their chief asset, powerful jaws, as weapons. Combat is not as deadly as it looks as the lizard's bite is strong but non-venomous.*
Bottom. *The pride and joy of a male fiddler crab is its huge, pink claw which is used to threaten other males.*
Right. *Locked in combat two male stag beetles make use of their 'antlers', actually over-developed mandibles.*
Page 40. *Each gannet pair in this huge flock sticks to its own small territory, an area they defend rigorously.*

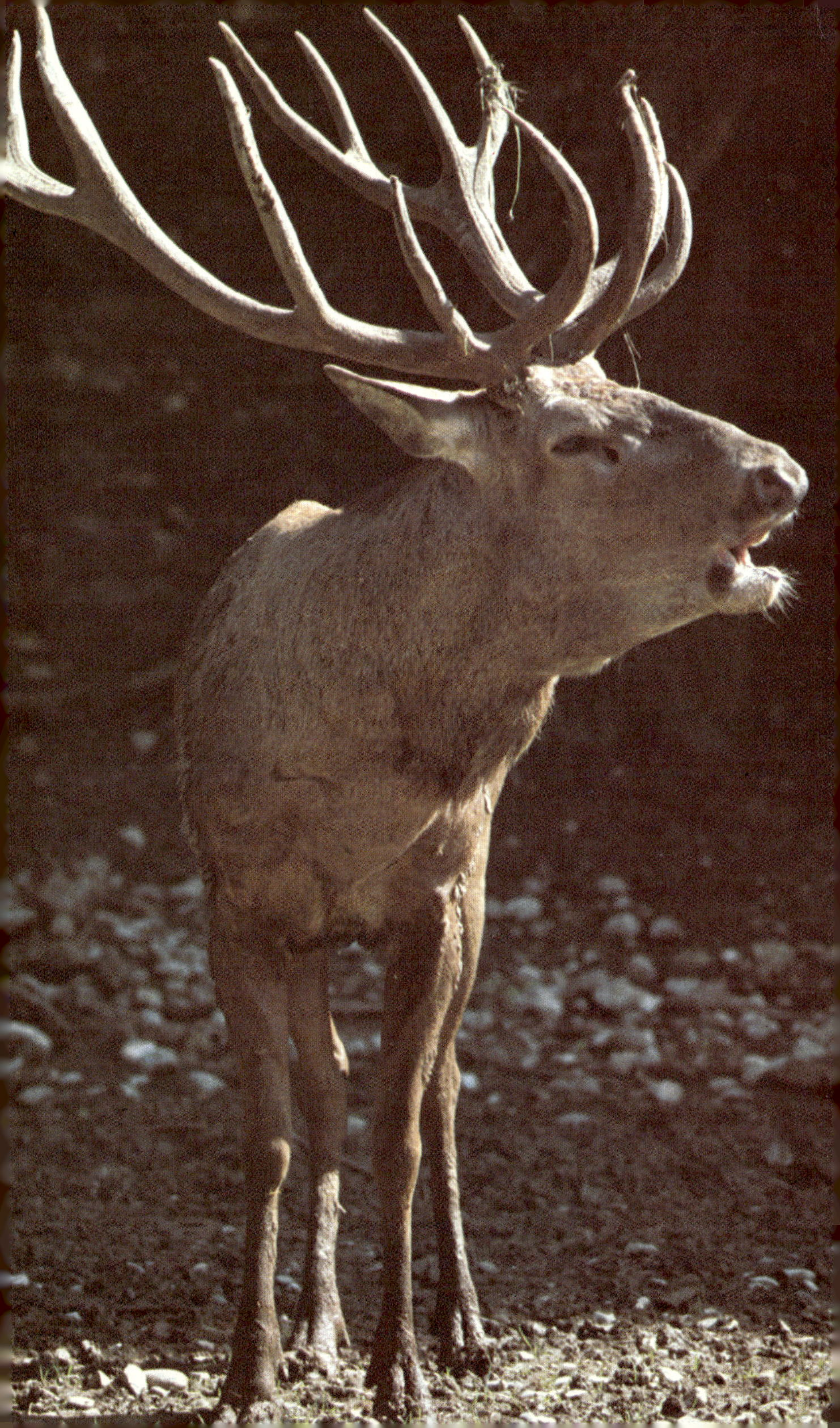

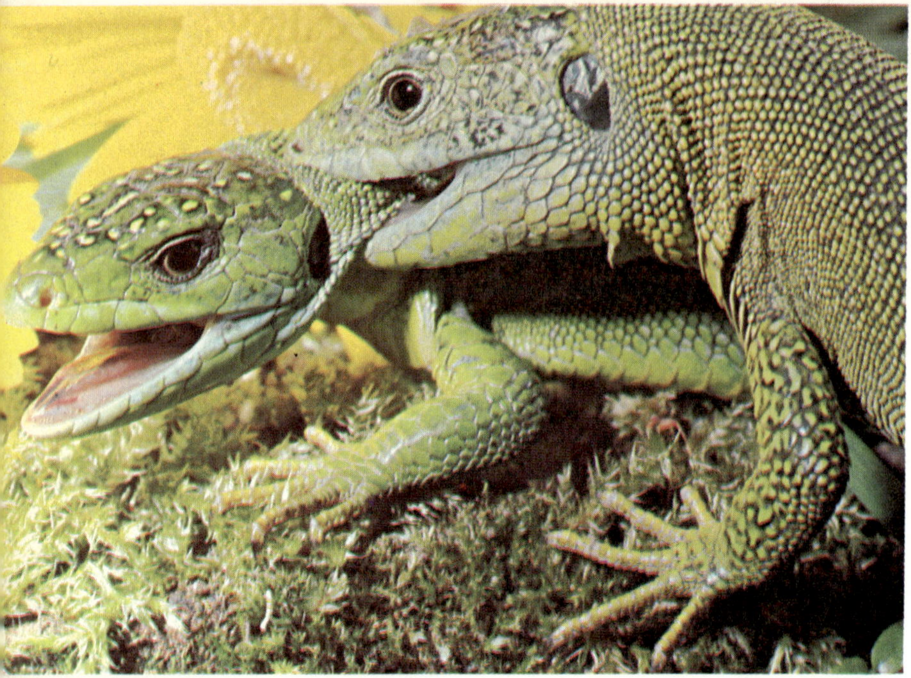

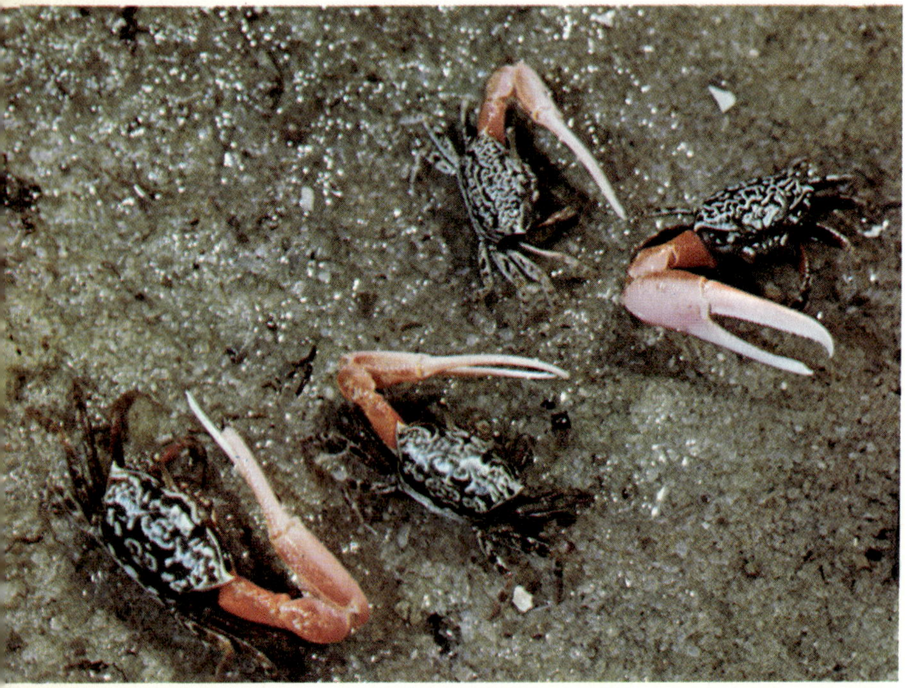

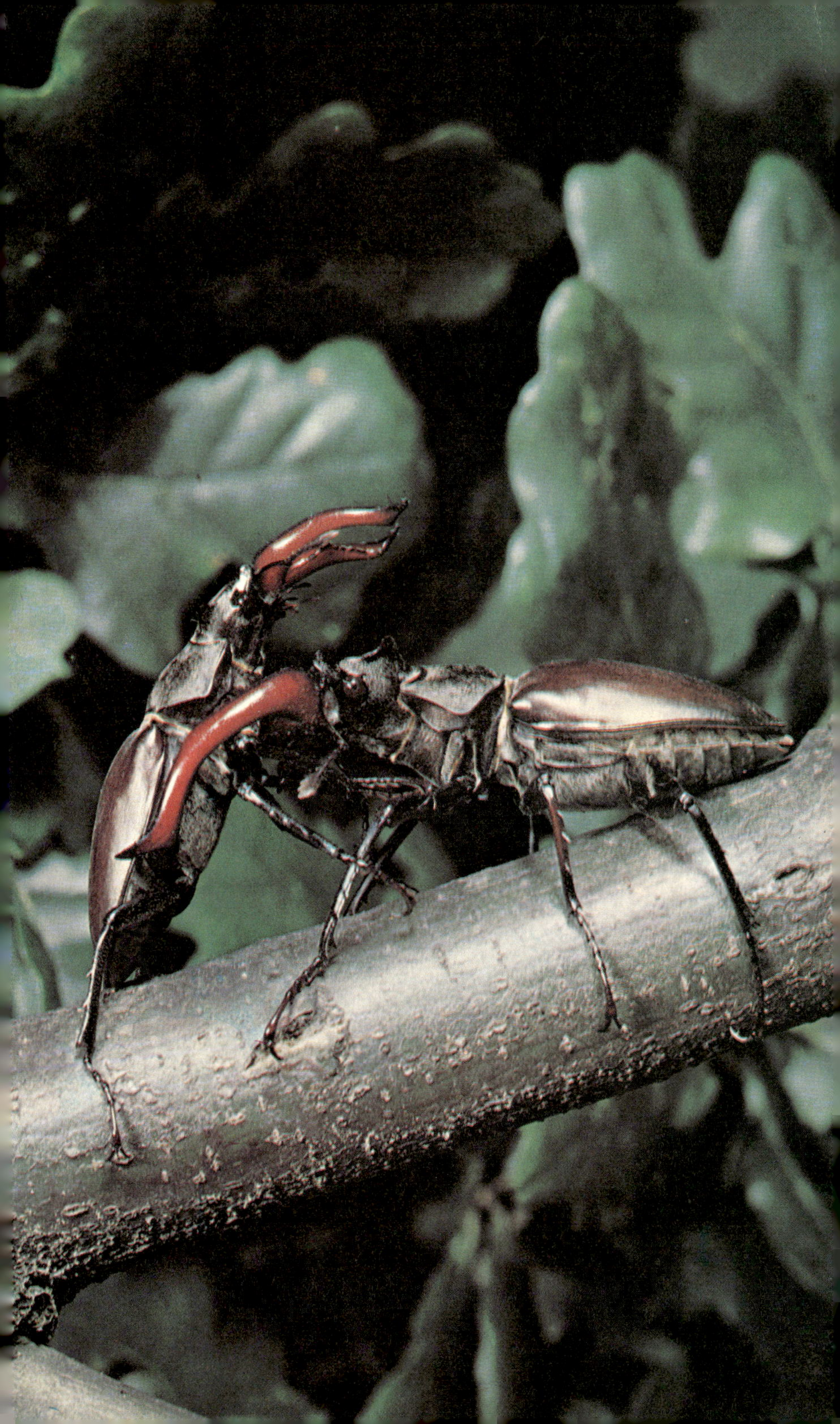

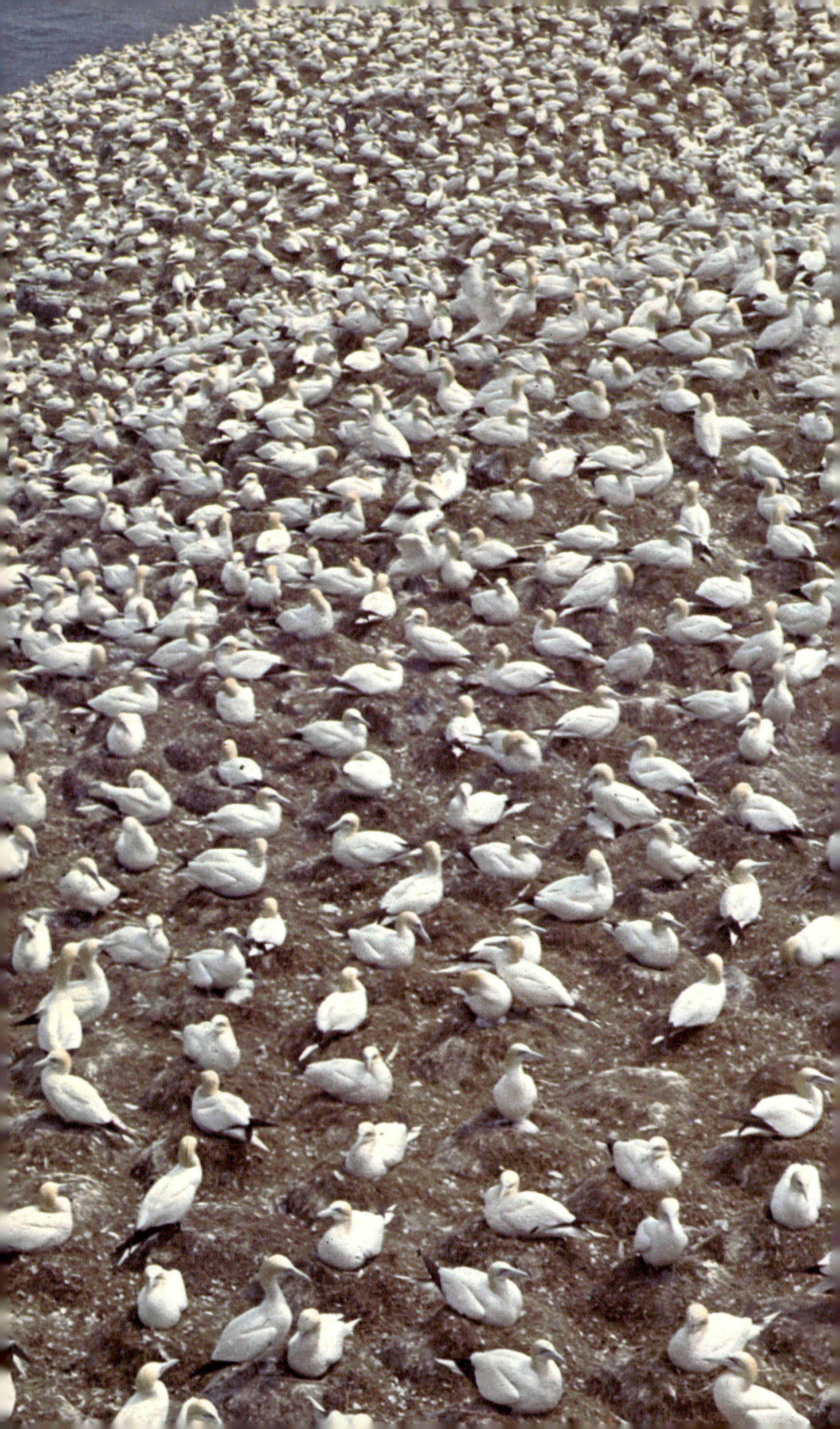

much bigger than he actually is and also quite terrifying. The intruder may now do one of two things. He either flees, in which case he is chased off the territory, or, more usually, he responds to the display by raising his own fins and becoming brighter in colour himself. The next step is that the two fishes circle one another trying to butt each other with the sharp edge of the jaw; they may even seize each other by the mouth in a trial of strength. In the end the intruder usually gives up and is finally chased away by the owner of the territory.

It is not only vertebrate animals that will defend their home and territory against intruders; there are some invertebrates that are territorial and readily take part in a fight. Fiddler crabs, which are found on tropical beaches all round the world, are well known for their territorial habits. They live in burrows in the sand, staying near their homes feeding on the small animals in the mud.

Male fiddler crabs are unusual in that they have one very large claw, the fiddle. They cannot feed with their fiddle but use it to signal from their burrows. This signalling is their way of marking their territory and of beckoning to females. This fiddle claw plays a large part in territorial disputes. If an approaching male disregards the signalling of the territorial owner it becomes more frantic until eventually fighting breaks out between the two crabs. They prance towards each other, challenging by patting the ground with their large fiddle claws. After lunging at each other a few times they finally interlock their large claws, a movement always cautiously taken as there is the possibility of the claw being damaged or broken off. With their claws well locked together they move backwards and forwards until the weaker of the two breaks away and retreats to its hole, its only injury being that of wounded dignity.

Among the insects, crickets are particularly noted for aggression. In mild encounters between two rival males they rear up the front half of their bodies and lash out with their antennae. If the contest is more serious these activities are followed by biting, head butting and attacking with the forelegs. But as always in intraspecific fighting mutilation is rare. Once crickets were specially bred, by the Chinese, for their fighting abilities. Fights were staged in special bowls, the end being when one cricket bit off the head of the other. Today this 'sport' is only played by children.

Dragonflies are unique in that they only keep their 'territories' for a day. In the morning the male looks for a suitable spot along the shore of a lake or river, about four yards in diameter, from which he can hunt his prey and also attack rival males flying past. If a female flies into this defended area and both dragonflies are 'in the mood' then mating will take place. The next day the male dragonfly will look for a new territory to defend.

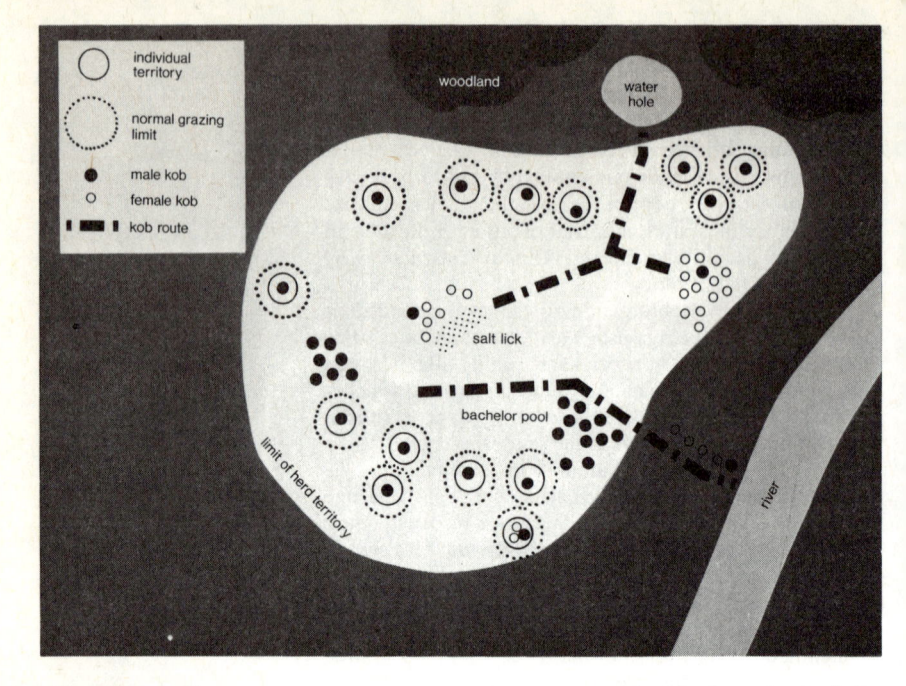

Claiming a home

Animals do not take part in fights any more than is necessary. When two rival males meet outside their territories they are not aggressive towards each other because fighting has no purpose in this context. An animal is more vulnerable to attack by predators when preoccupied by fighting, but this disadvantage is offset by the great value to the species of territorial activity.

The chief function of aggression is maintaining acceptable distances between animals, especially by means of territory. Once the boundaries have been established by fighting, there is a settling down and a reduction of strife. Most vertebrates – including fishes, lizards, birds and many mammals – and some invertebrates too, defend territory of one kind or another. But although the extreme importance of territorial activity is quite clear from the amount of time and energy devoted to it by animals, there is no complete agreement over the exact nature of the benefits which result.

The most widely accepted explanation is an economic one and concerns the availability of food. Animals benefit from being spaced out evenly over a particular habitat because their numbers are limited and competition for food is reduced. This can be very important in the breeding season, when the young may have to be left undefended while the parents forage. If food is near at hand, the period of absence is kept to a minimum. The territory

A restlessly grazing herd of African kob antelope is far more ordered than it may appear. The herd itself has a group territory, and within this a number of males hold roughly circular individual territories of 20–60 yd. diameter, often with a harem of females. Other bucks form bachelor groups or wander singly around the edges of the central area running quickly to a particular territory and challenging the occupant. Unattached females also form groups without territory and like the rest of the herd they follow regular routes within the herd's home area, such as those to the waterhole and salt lick indicated in this diagram of a typical kob herd territory.

itself may not be able to supply sufficient food to meet the needs of the owner, but it will probably be situated in a larger undefended area known as the home range, where the animal forages.

Defence of the young and avoidance of predation and disease are also aided by keeping up territorial boundaries. The possession of a territory means that each animal becomes very familiar with one particular spot. This familiarity probably helps the animal find food, escape from predators and increase its fighting potential. By the time a female arrives at the territory the male has usually established himself. The territory isolates each breeding pair of animals from the rest of the flock, shoal or herd, giving them freedom from disturbance during pair formation. It also helps seal the bond between the pair. When each male possesses his own defended area there is less likelihood of several males mating with one female. As each male actively defends his territory it makes sure that no two animals set up home in the same place. This would be disastrous for both families.

An animal's choice of territory depends mainly on inborn preferences possessed by all the members of that species. This means that all animals of a species select the same type of habitat. A male stickleback, for example, is born with the tendency to live in shallow water with plenty of vegetation, but he becomes conditioned to the landmarks of his selected territory. When these landmarks, such as pebbles or a particular plant, are moved, the stickleback will shift his territory accordingly. Each animal knows his own territory and that of others of his species by the landmarks.

For many mammals these landmarks are 'scent signals' deposited at the boundary and making use of natural features of the surroundings. A potential intruder is warned by them that he is trespassing. Dogs, foxes, wolves and jackals mark their territory with urine, sprinkling it on various objects such as trees and stones. The male bison marks a tree in his territory by barking a part of the trunk with his horns, then urinating on the ground nearby, rolling in the urine and finally rubbing his back against the barked tree so that the urine is rubbed off onto the trunk. Some mammals have evolved special scent glands just for the purpose of staking out their boundaries. The Indian blackbuck and other antelopes have such glands beside their eyes, which they rub against branches to leave a scent. The muntjac has a gland in its feet between the bases of its hooves as well as face glands.

Some animals advertise their territories vocally. The songs of male birds are not only to attract a female but also play a large part in marking a territory. In fact, the romantic myth that it is the female nightingale which sings is quite erroneous. The song which has delighted so many poets is that of the male as he advertises his rights over a

Territorial trademarks

1. *A Grant's gazelle 'marks' its territory by rubbing secretions from its face glands onto a bush.*
2. *Domestic animals also keep territories – this Alsatian marks a tree with urine which will advertise his presence to other dogs.*
3. *Between a muntjac deer's cleft hoof is a pedal gland the scent from which is used to mark territory. A cross section through the middle shows the indent where the gland is situated.*
4. *Most birds use song as a territorial marker. Here a corn bunting claims his boundary from a gate post.*

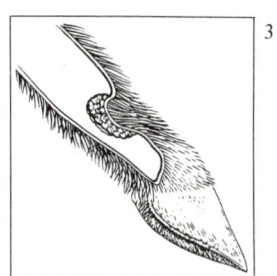

Fore and aft markers

5. *The male European bison has no scent glands. Instead it marks the tree by tearing the bark with its horns. After this it urinates on the ground, rolling in the liquid before returning to the tree and rubbing its back and sides against the trunk to impregnate the smell.*

6. *The rather weighty giant panda Chi-chi chooses a suitably sturdy tree upon which to rub her territorial scent signal. Pandas have special scent glands which are used as 'markers'. These glands are situated just underneath their short, stumpy tails.*

particular area. Boundaries are also psychologically fixed in the minds of animals, as we shall see in the next chapter, because they are associated with the changeover from success to defeat in disputes with neighbours.

In the majority of animals the defended territory is static, such as a piece of land, a tree, some waterweeds. There are a few animals, however, that choose to defend a moving territory. The male bitterling, a freshwater fish, closely guards a mussel in which the female lays her eggs. The mussel is not a sedentary animal but wanders around, although quite slowly. Many deer, and a waterbird, the avocet, confine their fighting to the vicinity of their chosen female wherever she may happen to go. This type of 'free' fighting can be thought of as defence of a mobile territory.

Choosing a mate

It is often said that rival fighting among animals makes sure that the strongest and healthiest males win possession of the female or females, and so become the fathers of the next generation. Probably the strength of the father does affect the welfare of the young, especially in a species like the lion where he plays an active part in care and defence. But it would not do to limit mating to a very few males. This would lead to serious inbreeding and a lack of variety in the inherited characteristics of the animals. Variation helps the species to adapt to new conditions. Sometimes the less dominant or defeated males are able to mate with the females while the chief rivals are fighting. For animals such as antelope, deer and bison that are not territorial but wander nomadically, rival fighting has another important function. It is a means of making sure that the strongest and most courageous males defend the families and herds. At every threat to the community the victorious males will immediately surround and protect the weaker members of the herd.

When the great evolutionist Darwin spoke of the 'struggle for existence', he was not only referring to the struggle between animals of different species, as seen in the prey and predator situation. What he really meant was the competition between animals and their near relations, that is between animals of the same species. Intraspecific fighting plays a useful part in the life of animals. The behaviour and structures used are highly developed, so they are obviously necessary and have evolved because they help to preserve the species.

Use as weapons in rival fighting may not be the only reason for the development of horns and antlers, however. The horns of some sheep, for example, are not of very great value as weapons, being wrongly shaped for fighting. They may be the result of an aspect of evolution called by

Right. *The tournaments of ruffs take place on special grounds or 'hills', where males (ruffs) display their fighting prowess while the females (reeves) look on. In a flurry of wings two males surge into airborne combat as another shows off his neck feathers or ruff.*

Below. *Though anatomically not very accurate, this old drawing of a displaying male argus pheasant clearly shows the beautiful but unpractical wing feathers.*

Darwin 'sexual selection'. It acts through the competition of male animals for mates. Sexual selection may explain why many male animals have brilliant plumage or large antlers or some other conspicuous feature absent from the female.

When a male is fighting or parading up and down in front of a rival, there are usually females present which, although they may not appear to be taking any notice, are affected by the male's behaviour. In some way they are probably stimulated and become more ready for mating. A male whose display impresses females more strongly than another's is likely to win more mates and therefore father more offspring. Some special feature of his body may be the cause of the stronger stimulus and may well be inherited by the young. During the course of evolution, the horns, feathers or other attributes concerned gradually become more and more showy and elaborate.

The bright colours of male pheasants, which make them conspicuous to enemies but more desirable to females, have probably been sexually selected in this way. Greater advantage to the species is gained if males are colourful and mate with many females before meeting an early death at the hands of a predator, than if the males are drab-coloured, for then they may live much longer yet still father fewer offspring.

The male bird of paradise and mandarin duck are two other birds which are beautifully adorned with bright, colourful feathers to attract the females. But perhaps the most extreme example is the argus pheasant. The hen pheasant is particularly responsive to the eyespot decorations on the secondary wing feathers of the cock. These feathers are so huge that the cock can scarcely fly, but the bigger they are the more aroused is a female. So the longer his decorated feathers become the more progeny the pheasant is likely to produce, and the less he is able to escape predators by flight. A more developed version of the plight of the common pheasant, the argus' situation shows how individual advantage is often sacrificed to the more lasting advantage to the species.

Sometimes female animals are said to choose their mates – favouring the most attractive suitor – rather than males fighting for females. There is not much difference however; it is just another way of saying that females mate with the males which arouse them most. When male ruffs hold their 'tournaments' of fight and display, the females act as spectators and seem to pick out the dominant male as the most desirable mate. He is the most victorious in battle and the most handsome in appearance. This choice suggests that sexual selection has been responsible for the evolution of the male's distinctive ruff of feathers.

Whether the male fights for the female or the female chooses the male, courtship is a time of stress for both of them. Before mating can occur the female must be sexually responsive and overcome her fear of the male and become used to the closeness of his presence. The male must be able to suppress his aggressiveness towards the female as at first he sees her as an intruder into his domain. For men and women, too, courtship is an arduous business. Although the man does the courting it is the woman who makes the yes or no decision!

Mass movement

A huge migrating herd of caribou moves across the Canadian tundra, completely lacking in any kind of internal organisation or territory yet moving together in a fairly coherent group. In the rut, males do not defend harems but there is some rivalry between them resulting in mild sparring. So territorial activity in these animals is at a minimum.

Group life

The herding instinct and aggressive instinct are not so incompatible as they may seem at first sight. Some animals are territorial for certain months of the year and gregarious, wandering freely, at other times. In red deer the males associate with the females only at mating time. During the rest of the year the females and young form separate herds from the males and neither sex shows territorial behaviour. In the rut, the males enter the home range of the females and fight amongst themselves to establish territories.

Other species, for example frogs, lack intraspecific aggression altogether, and do not hold territories, but nevertheless manage to avoid each other. Tree frogs live solitary lives except at the mating time and are obviously very evenly distributed over one particular area. This is achieved by each frog avoiding the quacking sound of its own species. But this does not explain the even distribution of the females, for in most frogs they are voiceless.

Many hoofed mammals as well as birds and fish live peacefully in large groups – herds, flocks and shoals – for most of the year. The advantages of a community life, defence against predators, warning each other in time of danger and other co-operative actions, are more important to these animals than keeping territories. They are also able to move to new feeding grounds to keep from going hungry. Only in the breeding season is the calm disturbed by rivalry.

In contrast, a kind of spacing out by means of aggression can be seen all the year round in flocks of birds such as chaffinches. Individuals are not tolerated nearer to each other than a certain minimum distance. The squabbles

Seabird suburb: rows of meticulously spaced gannet nests look uncannily like the monotonous streets of a human town. Each pair occupies and defends a small breeding territory centred on the nest, and is able to locate its own nest with great precision on returning after an absence.

seen in such flocks in winter are usually concerned with keeping up the spacing between each bird and its neighbour. There are undertones of male rivalry because a male bird allows females to approach closer than males. Starlings often sit like a string of beads along a telegraph wire, each bird at an equal distance from the next. The space between two birds is the exact distance each starling can reach to peck the other with its beak. If a starling arrives and joins the others but is too close to its neighbour, then the two birds peck at each other until properly spaced out. In effect each bird has a small moveable territory.

Within the huge nesting colonies of seabirds such as gulls, kittiwakes, terns and penguins, each bird's nest is surrounded by a territory of a few square yards. These enormously sociable birds still maintain their distance from each other. The normal function of a territory in providing a certain food supply seems not to apply to these birds. They do not get their food from their territories, but live off shoals of fish in the sea nearby. So other reasons for the spacing out have been suggested.

The rearing of the young without interference from other birds is one reason. Alternatively, it may be that the spacing out is a compromise. The eggs and chicks of herring gulls, for instance, are well camouflaged like the young of solitary ground-nesting birds, their colours blending perfectly with those of the surroundings. Marauding foxes and dogs find such prey only after much

A honeybee has entered a strange hive and is at once detected by the resident workers because of her foreign scent. As long as the intruder is suitably submissive, she will not be attacked and may even be accepted into the colony. But if an attempt is made to escape, an attack follows.

searching if the nests are widely distributed. If they are too close together, the advantage of camouflage is lost because the large numbers of eggs and young represent a rich and tempting source of food for predators. Once one nest is found the others are easily discovered.

On the other hand, social nesting has the advantages of mutual warning and communal attack. Adult birds can jointly attack the fox as it looks for eggs and chicks to eat. This may not drive the fox away but the effect is to interfere with its search for food. In addition, living in large communities means that as many birds as possible can exploit a source of food when it is available. If one bird spots a shoal, the others are soon attracted to it. The herring gull's territorial habits may represent the balance of advantage between the solitary and social life.

Other animals which live in groups have forsaken all claim to personal territory but recognise an area defended by the community as a whole. When a group holds a territory in this way, it may advertise the boundaries in just the same way as individual animals do. Howler monkeys of South America are so called because of their call, a loud howl, with which they proclaim their territorial rights.

Colonies of social insects represent this group territory taken to its greatest development. There is virtually no aggression within the colony – all is directed outwards at members of other colonies with which there is competition for food and living space. As members of one community mutually interchange food, the insects acquire a communal smell recognised by all other members. Insects from other colonies are recognised by their alien scent and may be driven away or attacked. A honeybee entering the wrong hive will be attacked and stung to death by the hive occupants, unless it takes up an attitude of submission.

An ant that has been foraging away from the nest may quite easily pick up other odours, which will be detected when the ant returns to the nest and is subjected to licking by the other ants. There will be very little aggression towards a genuine member of the colony because the strange smells will be only superficial. However, an intruder from another colony comes up against much more aggression. Attack takes the form of short, sharp rushes from a member of the colony, followed by gripping and biting with its mandibles. After such an aggressive assault the intruder usually runs away if still able to do so.

Competition between colonies of insects is particularly interesting because each group seems to treat the other as if it were a different species and therefore 'fair game' for attack. Something of the kind happens in the human species, when societies fragment into groups which distrust and misunderstand each other. When there is a real breakdown of communication, the danger of war is greatly increased.

The rules and regulations

In mammals, birds, reptiles, fish, insects, crustaceans, even in molluscs, species are known in which the reproductive season brings much intraspecific strife. Yet it is usually bluff rather than fighting.

from The Herring Gull's World by Niko Tinbergen

Fierce faces, snarling mouths, horrific screams – all these aggressive and usually frightening actions are common threat tactics used to warn off opponents. They form part of the animals' 'bluff' war where victory is gained by visual and vocal threat actions without battle or bloodshed taking place.

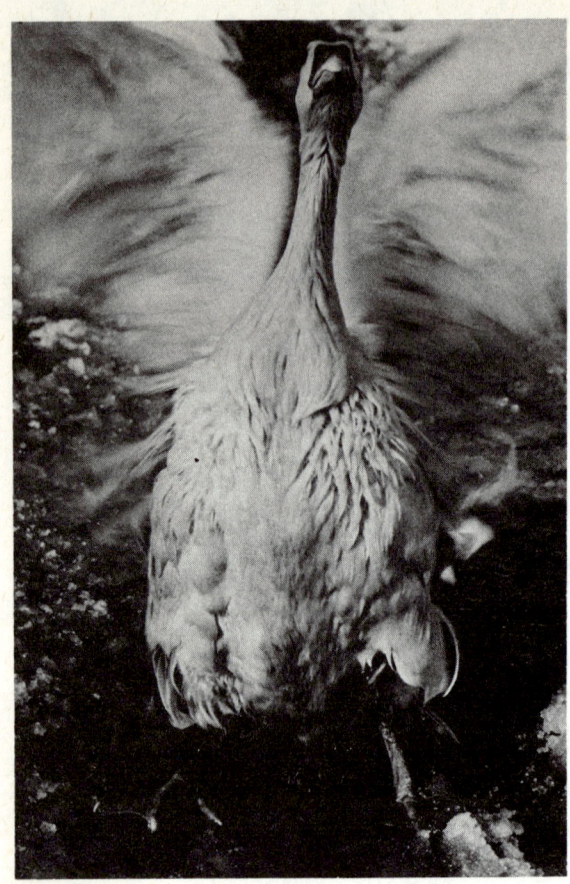

Fierce faces

Left. *An angry swan warns off an intruder. This wing-beating threat display is a form of bluff fighting.*

Right. *In an attempt to intimidate a rival male a Hamadryad baboon bares its large canine teeth.*

The study of behaviour is a study of what animals do and why they do it. As a science, this study had its origins early in the present century, since when research has continued along two rather different lines.

Following the example of the pioneer Austrian behaviourist Professor Konrad Lorenz, many scientists study animals undisturbed in their natural surroundings. They believe that an animal's behaviour can only make sense when observed in the natural environment of the species, and that it is not truly revealed in the laboratory. This is called the ethological approach. Other scientists prefer to experiment with animals under the controlled conditions of the laboratory, viewing behaviour from a psychological standpoint. Dr. Skinner of Harvard University, U.S.A. is famous for his scientific tests on behaviour in rats and pigeons performed in this manner.

Together the findings of these two schools of thought help us to understand just what is going on when two animals threaten or display to each other.

First of all, one must be aware of the source of an

animal's actions. Man is probably the only creature to have true freedom of choice. He can make up his own mind in a logical way when faced with a problem. In warfare, for instance, he can work out the most favourable positions to fit different situations. Animals, on the other hand, are much more strongly bound by fixed ways of behaving. They do not have conscious thought or the power to make a deliberate choice. Their decisions are, as it were, made for them. This is far from being a disadvantage, for when there is a choice it is easy to make a wrong decision.

Behaviour in its simplest form consists of the automatic reaction to some aspect of the surroundings – such as the appearance or actions of another animal – known as the stimulus. A stimulus is said to elicit a response. For example, the stimulus of a hovering bird of prey will probably cause a rabbit to respond by running for cover. The stimulus may come from within the animal too, as when hunger (perhaps in the form of an empty stomach) stimulates a search for food.

The response depends on the nature of the stimulus, the internal state of the animal, and what it has learned about the situation from past experience. An animal's age may also have a bearing on the response. Newly-hatched farmyard chicks will cringe with fright when any kind of bird flies by, but as they grow older they get used to certain shapes and no longer take any notice of them. Shapes which they see only rarely, such as a hawk hunting for food, continue to elicit the fear response.

Some of the more lowly animals, such as insects, live almost entirely by instinctive behaviour patterns. That is, their behaviour in fighting or any other context is largely unlearned. It is built into the nervous system and inherited just as body shape or colour are inherited. The caterpillars of the various species of processionary moths give an amusing and extraordinary example of instinctive

Head to tail, processionary moth caterpillars make their way to their feeding ground. This peculiar 'follow-my-leader' habit is unlearned and is a type of instinctive behaviour.

Opposite. *Iridescent eye spots in the peacock's train shimmer brilliantly as he begins his ritual courtship.*
Overleaf; left top. *A mixed herd of zebra and gnu share a territory, fights only occur at mating times.*
Below. *To serenade a female the one-inch long reedfrog blows up a vocal pouch almost as big as himself.*
Right top. *Affectionate puffins indulge in 'billing', part of the ritual overtures before mating takes place.*
Below. *Fine feathers on a mandarin duck may attract a female but have meant the loss of power in flight.*
Page 60. *An angry chameleon prefers bluff to blows and so threatens with open mouth and inflated body.*

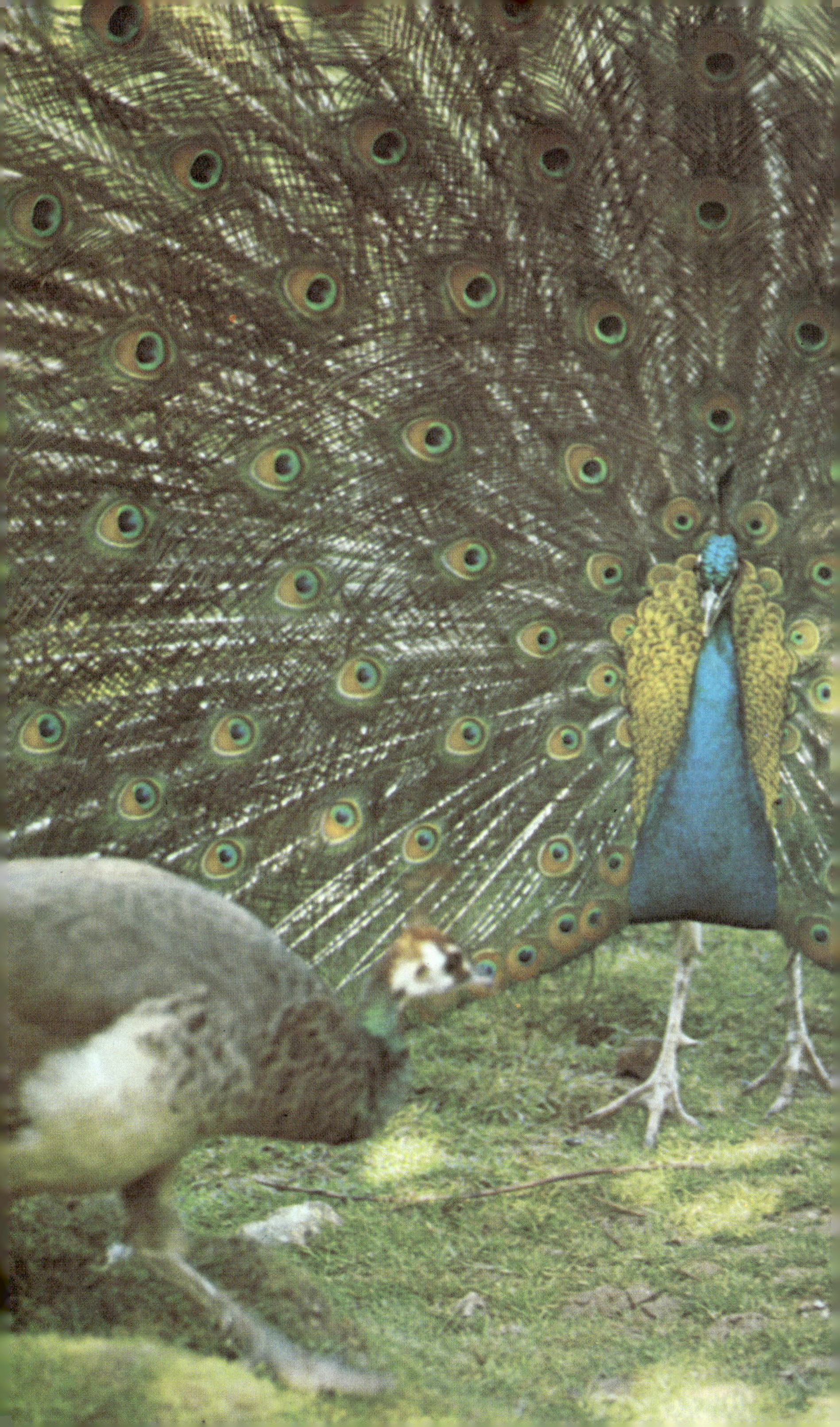

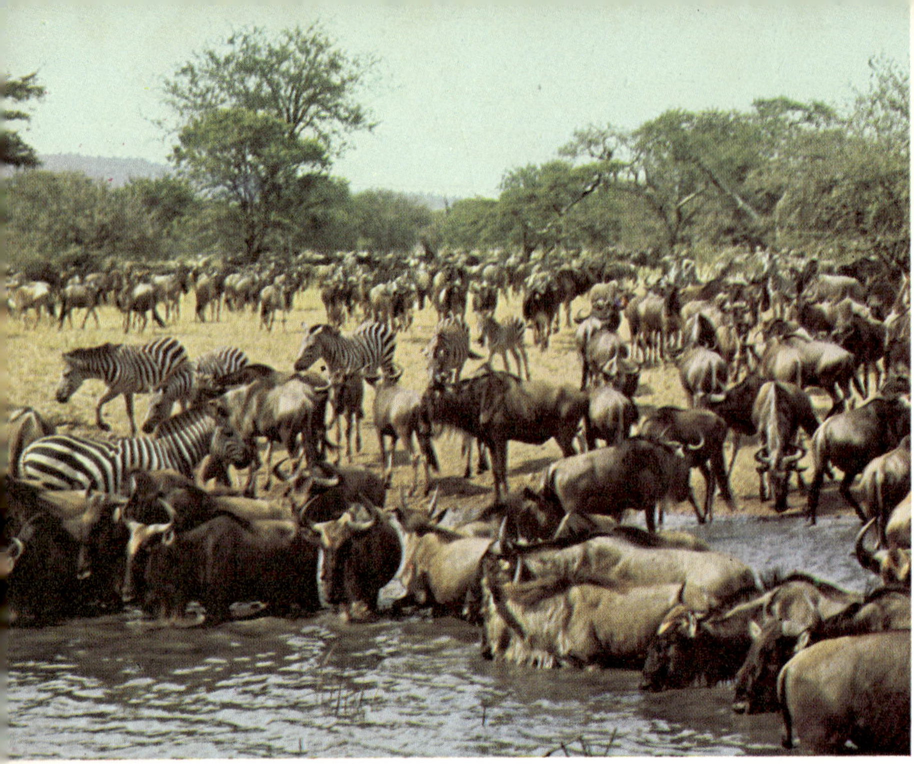

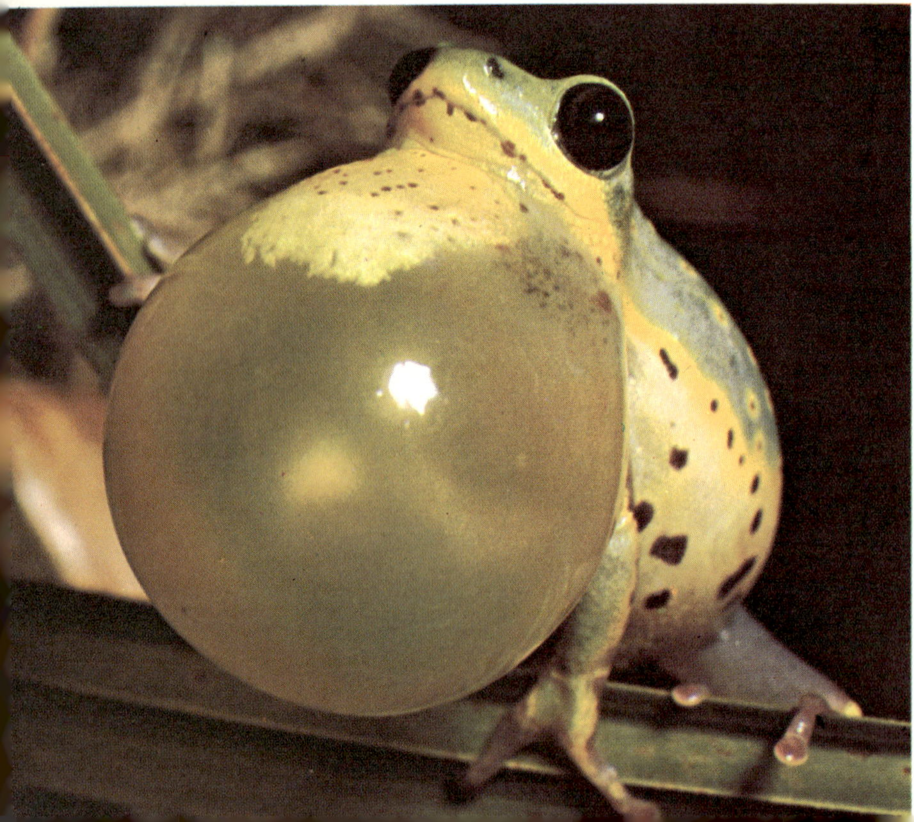

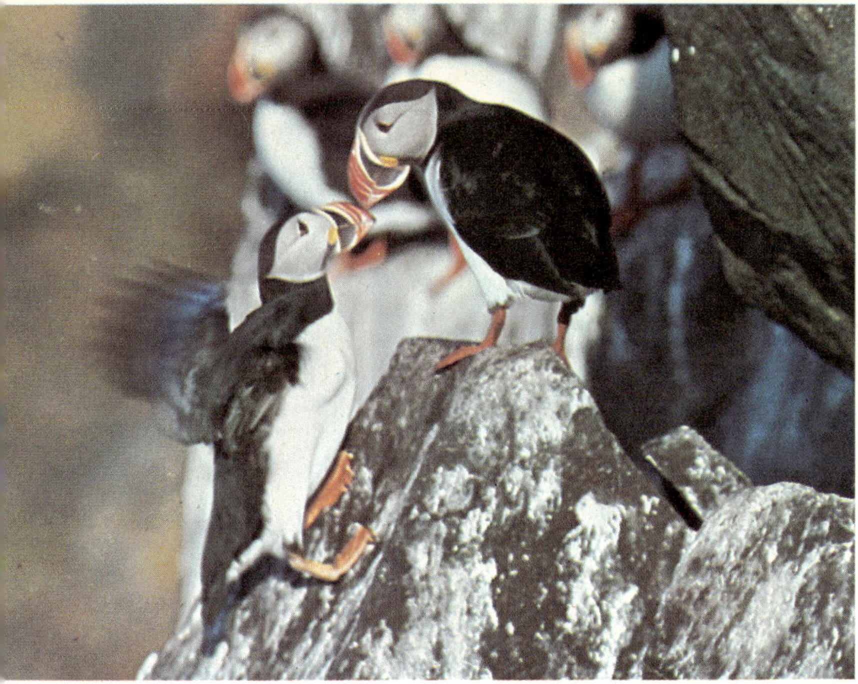

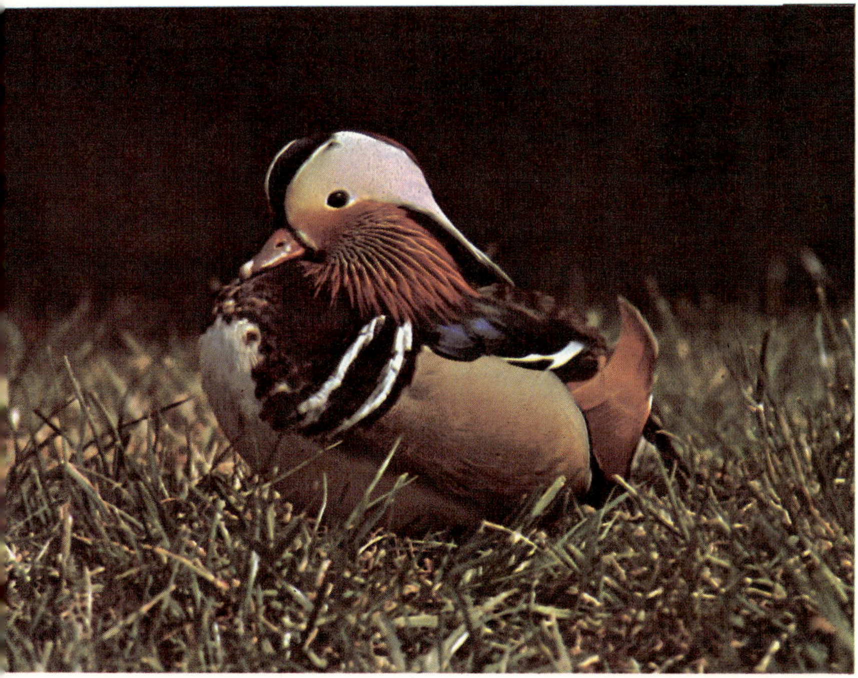

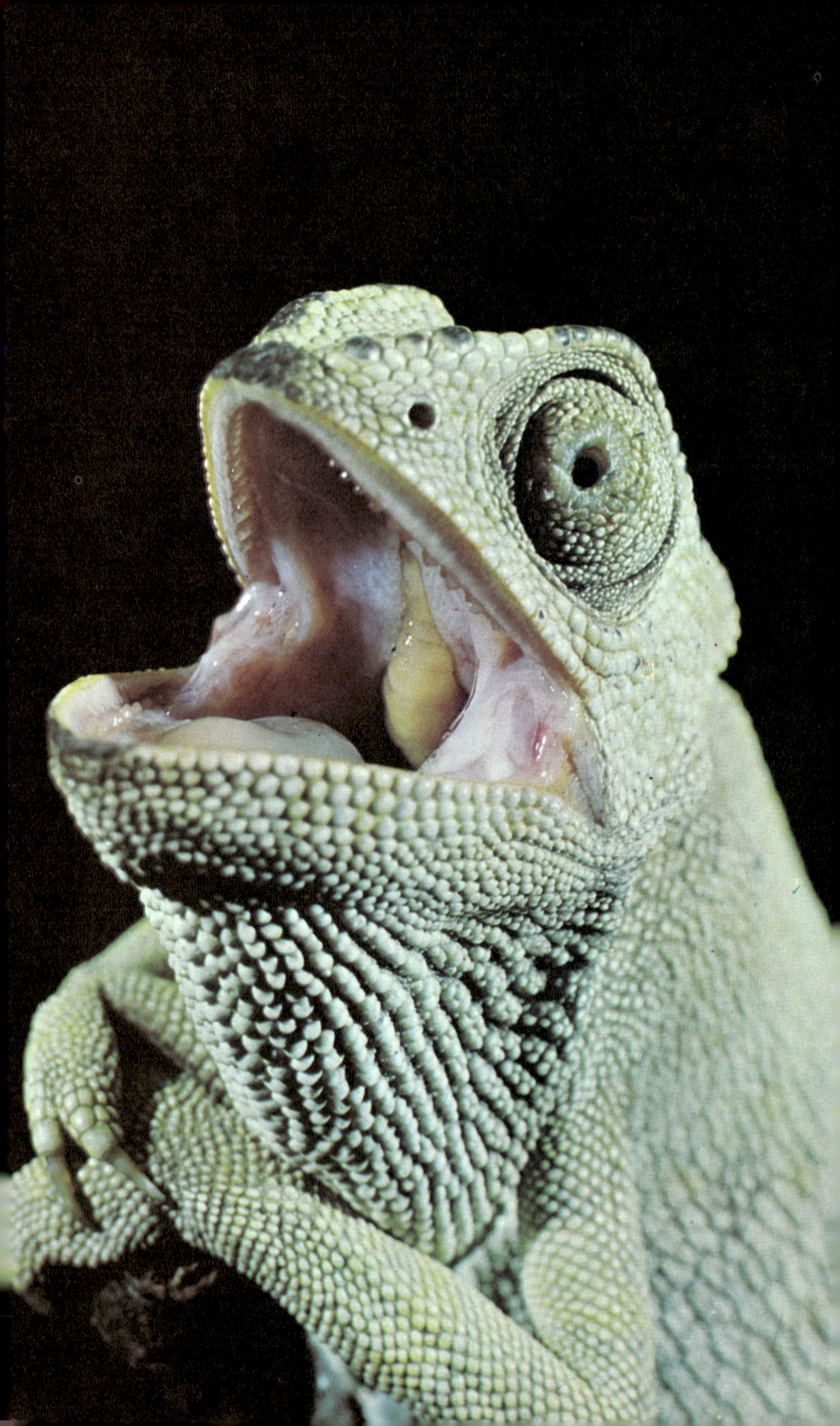

behaviour. Many caterpillars, up to three hundred, will gather in a long procession, each pushing the rear end of the one in front with its head. It might seem that this 'follow-my-leader' habit would lay the caterpillars open to dangers greater than usual. But one thing is certain and that is all the caterpillars are sure of reaching their feeding grounds and of returning to the security of their communal web after feeding. In higher animals, behaviour patterns are partly determined by learning, which is the capacity to modify behaviour in the light of experience. But unlike in man, the learning process is not a conscious one.

Unlearned behaviour represents a sort of 'species memory'. As a product of evolution it ensures that animals follow a course of action which has been proved to work and perfected over many generations. Thus behaviour becomes fitted to the normal environment. When there is a change in conditions, instinctive behaviour can also change but only slowly, through evolution. When an animal's actions are based on experience and learning, the animal is able to adapt much more quickly to the change.

Automatic responses play a very large part in every animal's life, especially during fights within a species over territory or mate. As we have said, rivals rarely engage in out and out battles but display threatening signals instead. Each species has its own battle code recognised and responded to only by its own kind, and this prevents time being wasted in fighting other species. The various types of threat signal are based on automatic responses to stimuli, responses which have often been repeated so many times that they have become stereotyped or ritualised.

Torn between two goals

The origins of threat are most interesting. It seems that man is not the only creature to suffer from frustration. Animals may also be thwarted or torn between two opposing aims. This is most likely to happen in the context of territory, when the tendencies to attack and to escape become mixed. Mating and satisfying hunger are two other aims which can prove mutually incompatible.

The escape tendency is sometimes called fear, though it is impossible to tell if animals experience this emotion in the same way as man does. Escape is a necessary counterbalance to attack, because in its absence there would be so much fighting that no time would be left for other activities.

The owner of a territory has confidence while on his home ground; he is in the right and knows it. An intruder is correspondingly unsure of himself because he is away from his own territory and trespassing in another's. At the very centre of his domain the owner has the complete psychological advantage and his attitude to a rival is one of almost pure aggression; he is responding to stimuli from the surroundings associated with success in former disputes. Nearer to the boundary, however, he is not so self-confident.

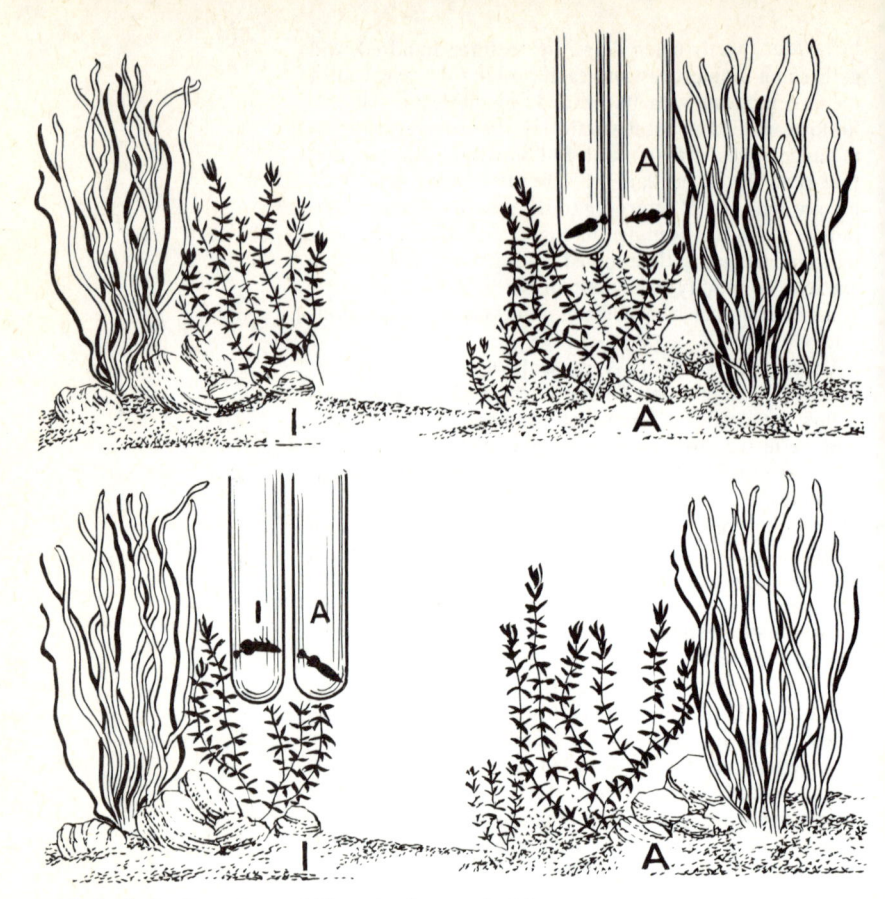

Boundaries between two territories are established in the first place by to-and-fro fights in which first one and then the other rival has the upper hand. Near the centre of his territory one animal will triumph but, chasing the other off the premises he becomes more and more uncertain. The further onto foreign ground he ventures the more likely he is to be the loser and to be chased away in his turn. After fighting for a while a point is recognised between the two territories at which the rivals are evenly matched. Here the tendencies to attack and escape are evenly balanced and the result of these mixed feelings is that both animals show threatening behaviour rather than actual fighting.

The psychological effect of territory can be shown quite easily in a simple experiment. Aquarium fish, especially

A simple experiment with sticklebacks can show quite clearly the psychological effects of territory ownership. Two sticklebacks, Albert and Ivor, each set up home, A and I, in an aquarium. By capturing the fish in test tubes it is possible to transfer them from one home to another. When Ivor is put in Albert's territory, top picture, Albert attacks with confidence, while Ivor tries to flee. If the situation is reversed, bottom picture, then Ivor becomes the attacker, his confidence restored as he is on home ground, while Albert tries to escape.

sticklebacks, make excellent subjects for this experiment provided the aquarium is large enough for each fish to acquire a territory.

Two male sticklebacks, let us call them Albert and Ivor, have each claimed their own little domain in the aquarium. They engage in rival fights if one of them trespasses on the other's home ground, but on the whole each fish knows his own boundaries and stays within them. To demonstrate this fighting behaviour, the two fishes are captured and each is put into a test tube big enough to move about in. When both test tubes are lowered into Albert's territory, Ivor shows signs of trying to escape while Albert makes attacking movements against the glass. The situation is completely reversed when the tubes are placed in Ivor's territory. His self-confidence restored, Ivor tries to attack Albert, while Albert attempts to make a quick getaway. This shows how closely attack is associated with territorial ownership.

Shown in their faces

When two tendencies clash, the result is called conflict behaviour. The simplest explanation for threat is that it is the result of internal conflict between attack and escape responses. The effects of the conflict can often be seen in the appearance of a threatening animal. In particular, the inward struggle may show in a mammal's face.

Anybody who owns a pet dog must have noticed his facial expressions when, on meeting a rival, he is torn between the desires to fight and to escape. His face may go through a number of changes, each one conveying the degree to which fear and aggression are mixed. An expression of threat is the result when the tendencies to attack and run away inhibit each other. It only needs careful observations to understand these expressions; just by watching his face it is possible to see the sort of mood

With ears well forward and teeth bared two huskies fight aggressively in the snow. Their facial expressions are similar to illustration C overleaf.

he is in. There is a gradation of expressions – shown in the drawings – depending on how strongly felt the aggression and fear are and how evenly balanced. A threatening face may show how fear of the rival is outweighed by the desire to bite him, but actual attack is still inhibited. Or fear may predominate with only a little aggression. Or both feelings may be equally expressed. With increasing fear the dog draws his ears and the corners of his mouth downwards and backwards, while with increasing aggression he lifts his upper lip and opens his mouth ready to bite.

The series shows the transition between the dog's fight and flight expressions. We can see that dog A is on the alert but does not yet know what action to take. The dog I is in a state of great tension with both fight and flight tendencies fully aroused; if the rival should move just one step towards him there will follow a desperate attack. This is an extreme situation and probably arises only when a dog is cornered or trapped, or when a bitch is defending her young against an approaching enemy. More commonly seen are the expressions in between these extremes. With experience it is possible to deduce the dog's probable course of action from its expression.

When a dog meets a rival it expresses its mood of fear or anger with its face. In the diagram the fight impulse increases to the right and flight increases downwards. Dog A is on the alert but unsure what action to take. Dog C is unafraid of the rival and is ready to fight while Dog G draws its ears back in fear. Dog I is in a state of tension with both moods of flight and fight fully aroused. The intermediate Dogs B, D, F and H show mixed feelings of fear and anger. Dog F for instance shows more anger than fear, Dog H shows more fear than anger. The middle Dog E shows equal amounts of both fear and anger.

Preparing to attack

In the threat postures of the dog and other animals, behaviour appropriate to both attack and escape appears in a sort of compromise. It is as though, unable to decide whether to go on or turn back, the animal stays on one spot. At the same time it may make movements as if to flee or attack. These are called preparatory or intention movements and they arise as a direct response to a situation. When human beings shake their fists at each other the intention to strike is implied.

An animal example comes from the herring gull. These birds have been extensively studied in England by Professor Niko Tinbergen of Oxford University. The orderly arrangement of the birds' territories within the nesting colony is not always apparent, as there is much coming and going of gulls and persistent squabbling. If a male returns from a food forage and finds a stranger within his home, he walks rigidly up to the trespassing bird with his neck stretched upward and forward, and his head pointing downward. This is an exaggerated version of the posture used by a gull before delivering pecks at an opponent and is called the upright threat posture.

If the threatening bird feels particularly angry towards the intruder, he will lift his wings a little so that they stand out from his body and are ready for fighting. These intention movements of attack spell out to the rival exactly what mood the territory owner is in, but show that

The upright threat posture of the herring gull.

65

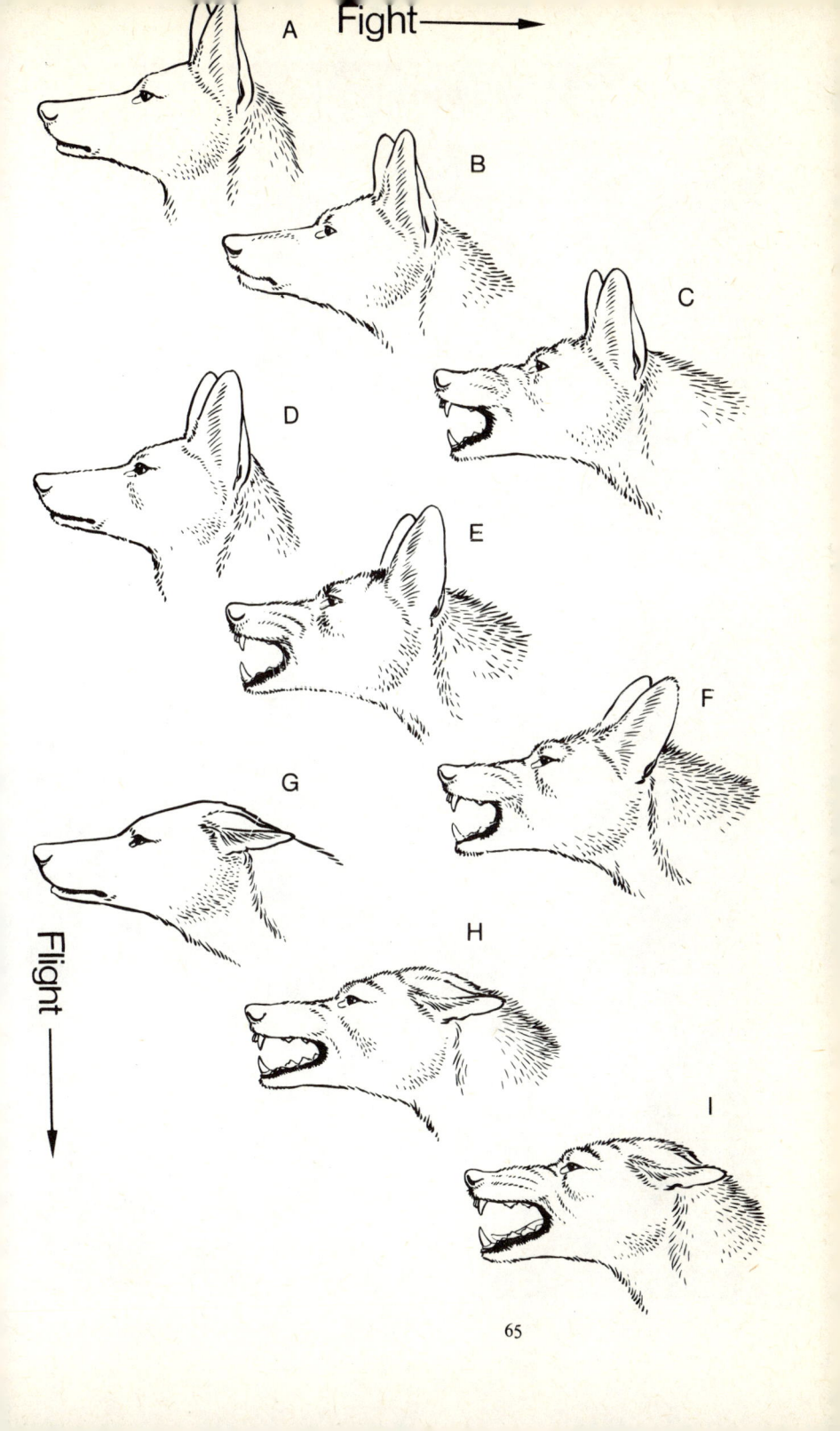

A matter of bluff

Left. *Inhospitable paper wasps take up a threatening attitude on their nest of papier mâché cells.*

Right. *To deter a rival the long-eared owl hunches its body up, opens its wings and fluffs out its feathers. The owl also raises its 'ears', long tufts of feathers on its head. All this has the desired effect of making the bird look twice its normal size and so very formidable.*

Below. *The hostile display of the bittern is similar to the owl's. It, too, spreads its wings in threat.*

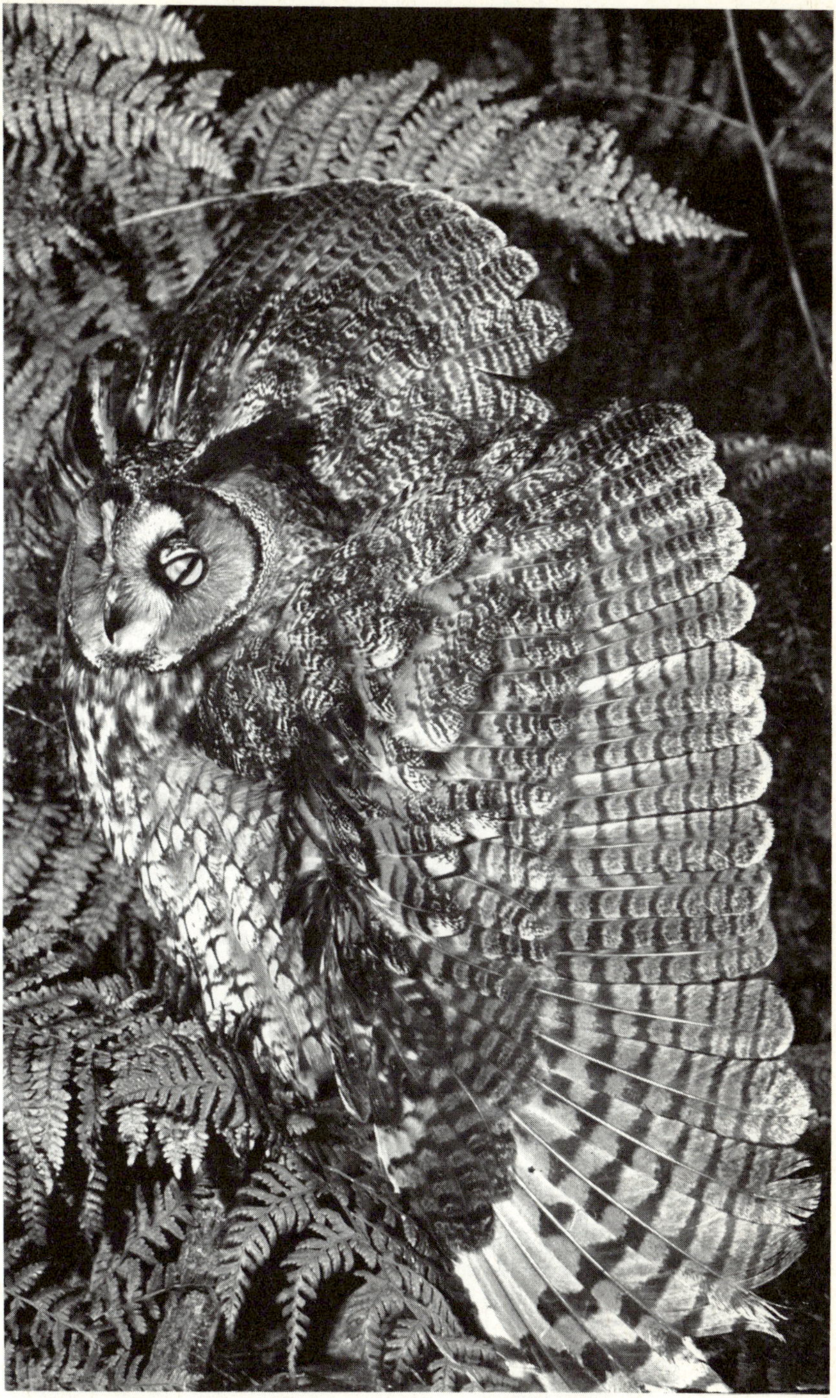

the gull's tendency to attack is not fully aroused and is mixed with some apprehension.

The intention implied in the upright threat posture is demonstrated when the resident male is provoked into further action by the persistence of the intruder. He lifts his wings higher and higher and finally charges towards the stranger, half running, half flying. This is usually enough to ensure retreat, but should the intruder be standing near ground that is familiar, he will run only just past the territory boundary. Then, his confidence restored, he will stop and raise his powerful wings to their full span and stand still for a few seconds, his neck stretched in the upright threat posture. The resident male will not let the one-time intruder have the last word, however, and he too takes up this position on his side of the boundary as a parting gesture of territorial rights.

Intention movements are usually obvious in gull behaviour and they clearly reveal what sort of activity is to be expected. An intruder, for instance, sometimes regrets his act of trespass and becomes frightened, though still slightly inclined to attack. When this happens he adopts a special anxiety posture – he stretches out his neck but keeps his head horizontal or slightly raised rather than pointing downwards. Throughout the territory owner's display the intruder never faces his opponent; he is

The anxiety posture of the herring gull.

Left. *Another displacement activity can be seen among sticklebacks in a ritualised form. When two males are feeling aggressive towards each other, one of them may sometimes make as if to dig into the sand, a gesture which looks like the start of nest-digging. It has in fact come to be a sign of threat.*

Far left. *Avocets sleep with the head 'tucked under the wing'. But this bird is not asleep. It has performed this displacement activity in the midst of a tense encounter with a rival, showing that the tendencies to flee and attack are both aroused at once, and neither can express itself.*

afraid and ready to flee and this is understood by the other gull.

When a bird is alarmed by an approaching disturbance it makes intention movements before actually flying away. It raises its wings, bends its heel-joints and stretches its body strongly in the direction of flight. Such movements are a preparation for flight if this should become necessary, and are interpreted as such by other members of the species.

Irrelevant actions

Another component of territorial display is displacement activity, that is, behaviour which does not arise directly from the situation and is unexpected because it appears to have little to do with actual combat.

It seems that attack and escape responses may be simultaneously elicited so that neither can find an outlet. As if to ease the tension, irrelevant actions such as mock feeding or grooming are performed. Displacement activities may occur suddenly while a fight is actually in progress. Starlings, cranes, terns and many other birds may start preening their feathers, while the oystercatcher, avocet and other waders often quite unaccountably turn their head round and tuck their bill under the wing as if to go to sleep. Fighting cocks may pick up imaginary food. Great tits and blue tits – tree-feeding birds – will tear buds apart on the branches where they are fighting.

Sticklebacks have a distinctive threat posture which they use against a rival. The fish suddenly points his head down and makes jerky movements as if to bore himself vertically into the sand, snout first. This head-down posture is also a displacement activity. It often goes so far as sand-digging, the first stage of nest-building. But the intruder knows that the actions are not just nest-digging because the owner turns his broad side towards the opponent and erects his ventral spines; such actions do not take place in otherwise normal sand-digging.

When two male sticklebacks are placed in a small aquarium so that their territories are too close together, a strained situation arises. Neither fish is able to chase the other away without infringing on the other's territory. The outcome is almost continual displacement sand-digging until the aquarium is littered with pits.

In mammals, where learning plays a much greater part in behaviour than instinct, displacement activities are less marked, but there are a few examples. A wounded elephant will suddenly push saplings over in apparent fury,

Flesh and feather display

Like something from another world, a male sage grouse puffs himself up in a display which makes his head almost disappear among the fluffed up feathers and inflated air sacs. Such displays are very impressive to other males and to females in courtship.

and fighting rats may groom themselves. Human beings scratch their heads when puzzled or drum their fingers when nervous.

Letting off steam

Rather similar to displacement activities are actions which have become redirected to some object other than the opponent. The grass-pulling action of the herring gull may be an example of a redirected movement. It tends to occur when two male gulls try to defend the same plot of land. The angry males adopt their battle postures and walk stiffly towards each other like soldiers on parade, until they are about a foot apart. Then one bird suddenly lunges at the ground, pecking vigorously and tearing out grass, moss or whatever happens to be available. He may hold it for a short time before tossing it away.

The grass-pulling action of the herring gull.

The gull may then repeat this performance and the opponent also takes part if he feels aggressive. One bird may dash towards the other trying to get hold of the grass in his rival's bill, pulling hard if he does manage to grip the bundle. Often a gull may try to get hold of some plants that are too firmly rooted, so that pulling with all his might he ends up by tumbling backwards and making a spectacle of himself instead of frightening his rival. To the casual observer it seems that the bits of grass are a coveted prize. As the gull tugs viciously at a clump of grass he is doing to the plant what he would do to his opponent if only he had the courage.

Another explanation for grass-pulling is that it is displacement nest-building, but it is quite obvious to other birds that the gull is not building a nest by the very aggressive way he pulls at the grass. Men and women often relieve tension by redirecting anger to some inanimate object. Thumping a table or stamping the feet is perhaps the least destructive action of this kind, but gives little satisfaction. Breaking china, especially valuable china, and throwing things, are much more fulfilling. In the same way animals let off steam with redirected aggression when they are not sure enough for physical attack but must make some positive action.

Display of finery

Yet another element in threat displays are the physiological side-effects of conflict. Just as our hair stands on end or we breathe faster when we are badly frightened, so animals experience similar involuntary changes. Inflation of air sacs or lifting of gill-covers indicating a greater air intake,

One of the most common aspects of threat is seen when an animal tries to make itself look more frightening than it really is. The frilled lizard of Australia does this by erecting a large frill around the head, so appearing suddenly to increase in size.

and erection of feathers, are changes of this kind which have become exaggerated because of their value in making the animal suddenly look more impressive and formidable than it really is.

North American sage grouse cocks display at each other by stretching out their tails into a fan, so that, viewed from the front, the head appears to be surrounded by a spiky halo. At the same time they expose the red wattles over each eye or blow out the air sacs on their throat. Certain lizards, for example the small American anole, inflate a throat sac during threat display.

Some species have developed special crests, manes or fringes which they erect during display to intimidate rivals. The male kagu, a rare bird found only on the island of New Caledonia, has a long shaggy crest which it raises like a fan when it is angry. The ear tufts and collar of feathers of the male ruff, and the crest of the European spoonbill are also used in threat.

Among the mammals, the maned wolf is named after the thick hair along the back of the neck and shoulders which it erects in moments of excitement. Similarly, a well-known feature of the domestic cat is that when aroused it erects the hairs along its hunched back and seems to prickle with rage.

The frilled lizard of Australia has a warning device

Choking position of male and female herring gulls.

akin to other animals which 'swell' with rage in some way. When threatening an intruder it unfurls an umbrella-like frill which is eight inches or more across, at the same time opening wide its mouth in a menacing gape; all in all an awesome exhibition.

Intention movements, displacement activity, redirection, physiological changes – all these innate behaviour patterns are aspects of conflict behaviour which combine in threat displays. Instead of remaining in their unmodified form, where their meaning might be ambiguous, each has become ritualised during evolution so that their combined effect is a display specific to each type of animal and having a recognised meaning to all concerned. The effectiveness of visual and other displays depends on how awesome and conspicuous the threatening animal becomes. With a premium on the most impressive performance, it is easy to see how evolution has caused the ritualisation and exaggeration of the 'by-products' of conflict behaviour. Their original function may be quite obscured and in many instances the derivation of a particular part of a display cannot be classed into a type of behaviour such as displacement activity. Indeed, very little at all is known about the origins of invertebrate displays.

The value of threat – which is really bluff fighting – is that it allows both contestants to get their war-like feelings out of their system, and as one opponent usually gives up and retreats there is victory without injury. Although we have spoken of animals putting on these displays to intimidate their rivals, it must always be remembered that the actions are automatic responses to the appearance or actions of the rival and are not consciously undertaken with a purpose.

A gull's mistake

The preening movements made by cranes in threat display are ritualised. The movements are rigid and formalised when used as a threat so as to be quite distinguishable from normal feather preening. But it does not necessarily follow that every social signal of this kind has become ritualised. The choking display of male and female herring gulls is not, as yet, a ritual. A female will sometimes help a male gull defend their territory. The two gulls will rush towards an intruding pair with lowered breast, legs bent and head pointed downwards. At the same time they make a peculiar facial expression by lowering the tongue-bones. The resident pair make jerking movements of the head as if to peck into the ground, although the beak never reaches the ground. This peculiar performance is called choking.

But in contrast to this hostile choking, a pair of birds often performs the choking behaviour at the site of their

Animal morse codes

Ears, eyes and tail are three indicators of mood. A confident wolf holds his tail high but signals worry by lowering it. An angry tiger flattens its ears to show the white markings and a mangabey exposes its white eyelids in threat.

future nest, even when there is no stranger present. The actions are very similar and misunderstandings sometimes occur. Professor Tinbergen witnessed an incident in which a male herring gull mistook a mated female's hostile choking for the friendly kind. The male approached the female while she was in her territory, but without her mate. The female did not walk over to him as she would have done had she been prepared to be friendly, but responded by making choking actions. The male, incorrectly interpreting these actions, walked on as if to join her when she suddenly rose to attack him.

Such an incident would never have happened had a distinction been made between the friendly and hostile gestures, for example by making the threat display a ritualised one. In laughing gulls, which have the same choking behaviour, such a distinction does exist. In hostile choking at the territory boundary the bird ruffles its body and neck feathers, whereas in friendly choking the feathers remain smooth against the body.

Laughing gulls take care not to confuse the friendly and hostile types of choking. The 'ruffled' gull is in an aggressive mood, while the other is friendly – his back feathers lie flat and smooth in their normal position.

Signalling devices

Displays often help to show off to its opponent a particular feature of an animal, such as special colour patches or a certain posture. Such features were called releasers by Professor Lorenz because they are social signalling devices which release a specific, predictable response (in this case reciprocal threat) in the rival.

The releasers of every animal species must by their very function have certain qualities. They must always be conspicuous – to be effective a signal must be eye-catching (or nose-catching or ear-catching). Because of this conspicuousness it also means an animal is more prone to attack from its predators. The signals must therefore be able to be put on show and withdrawn instantly. The erection of hair or feathers, the inflation of an air sac, such signals can be switched on and off at a moment's notice. The releasing signals must be simple to set off the correct reaction in the rival, and the releasers must be specific so that they evoke the right response from one species only. This means that each animal has its own secret battle code. All releasers are inborn and the reactions they release are likewise innate. Every animal understands perfectly, right from the start, the meaning of their signalling devices. To have to learn their meaning or to misinterpret them could be fatal.

Releasers need not necessarily be visual. The marking of territory boundaries by scent or sound mentioned in the preceding chapter act as threat signals. In the visual displays of animals, however, colour is an important releaser as it is a feature so quickly and easily recognised. During the breeding season the male three-spined stickleback becomes transformed from a greeny-brown, insignificant little fish to one that is brilliantly coloured. His belly becomes a bright red, his eyes prominently blue and his back acquires a bluish hue. These garish colours are related to territorial ownership and the aggressive mood of the male, and are involuntary changes like the hair erection described earlier. They have a very dramatic effect on other male sticklebacks, as the colours act as threatening signals warning off rivals.

The red belly is the important releaser in the visual display. This can be demonstrated quite easily by an experiment using coloured models. A model shaped exactly like a stickleback but coloured completely silver receives no aggressive response from a territorial owner when put within his territory. However, crude cigar-shaped or roughly oval models that are not a bit like a real fish but have their undersurfaces painted red, immediately trigger off threatening signals from the territorial owner. The shape and size of the models are irrelevant; the red undersurface is the vital factor in eliciting a threat response.

This emphasis on colour rather than shape to denote a rival can occasionally result in amusing incidents. Lorenz reported a stickleback reacting quite violently to red vans passing by a window near which the aquarium happened to be. The fish was prevented from making an 'attack' only by the glass of the tank.

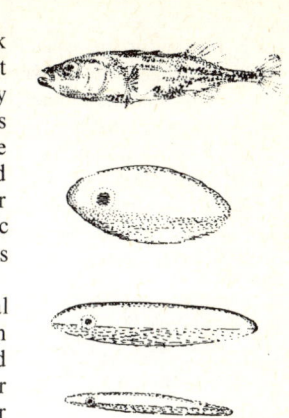

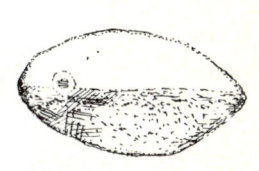

Although the top model looks to us far more like a real stickleback than the others, it does not evoke threat in another stickleback as do the cigar-shaped models. This is because only these models have red colouring, a releaser of threat behaviour in the stickleback.

Red, white and blue

Many other small animals are equally aggressive and are quick to react to colour stimuli. Robins have very strong territorial instincts. Both the males and the females hold territories for most of the year, so there is much boundary fighting. Like all birds the robin marks its territory with song and it is through song that a territory owner locates a challenger. Once found it is the red breast of the rival that arouses the threat display. Just as the stickleback will threaten any models that have red undersurfaces so will a robin display at a small bunch of red feathers that has been put within its territory. And just as the stickleback does not react to a perfectly shaped model that is silver, neither will a robin threaten an immature mounted bird that has a brown breast. It is quite remarkable that the red breast should have the same signalling function in two entirely different groups of animals.

Territorial birds do not usually attack intruders of other species unless they are predators, but sometimes there is misunderstanding, as when another red-breasted bird is unfortunate enough to happen upon an aggressive robin. Then occurs an example of inte*r*specific fighting which is really meant for in*tra*specific purposes. Like the stickleback's mistake with the vans, an occasional instance of this sort of mistaken identity does not matter, so long as rivals are *always* attacked. This is ensured by the system of releasers and the responses to them.

Male cuttlefish have a brilliant visual display in which colour plays a prominent part. On meeting another male, the cuttlefish shows the broad side of his arms and at the same time develops a conspicuous pattern of very dark

Red signals attack

Right. *It is the bright red breast feathers of a rival bird that arouses a territorial robin to threaten and attack.*
Left. *Even a stuffed robin put within an occupied territory is assaulted. A tuft of red feathers would have the same effect whereas a mounted immature robin, with brown breast feathers, would remain unscathed.*

Overleaf. *To intimidate rivals or enemies animals adopt many ways of appearing larger: gaudy colours are shown off by the firemouth (1) which spreads out fins and gill covers to expose its 'false eye' pattern; the frogmouth (2) opens its huge mouth and fluffs out its feathers; the boomslang (3) shows bright scales by inflating its neck and the cockatoo (4) displays by erecting its crest of long yellow feathers.*
Right. *A bristling wild cat threatens before attacking.*
Page 80. *A female cattle egret submits with lowered head to her mate's aggressively raised nuptial plumes. The third egret is merely an onlooker at this display.*

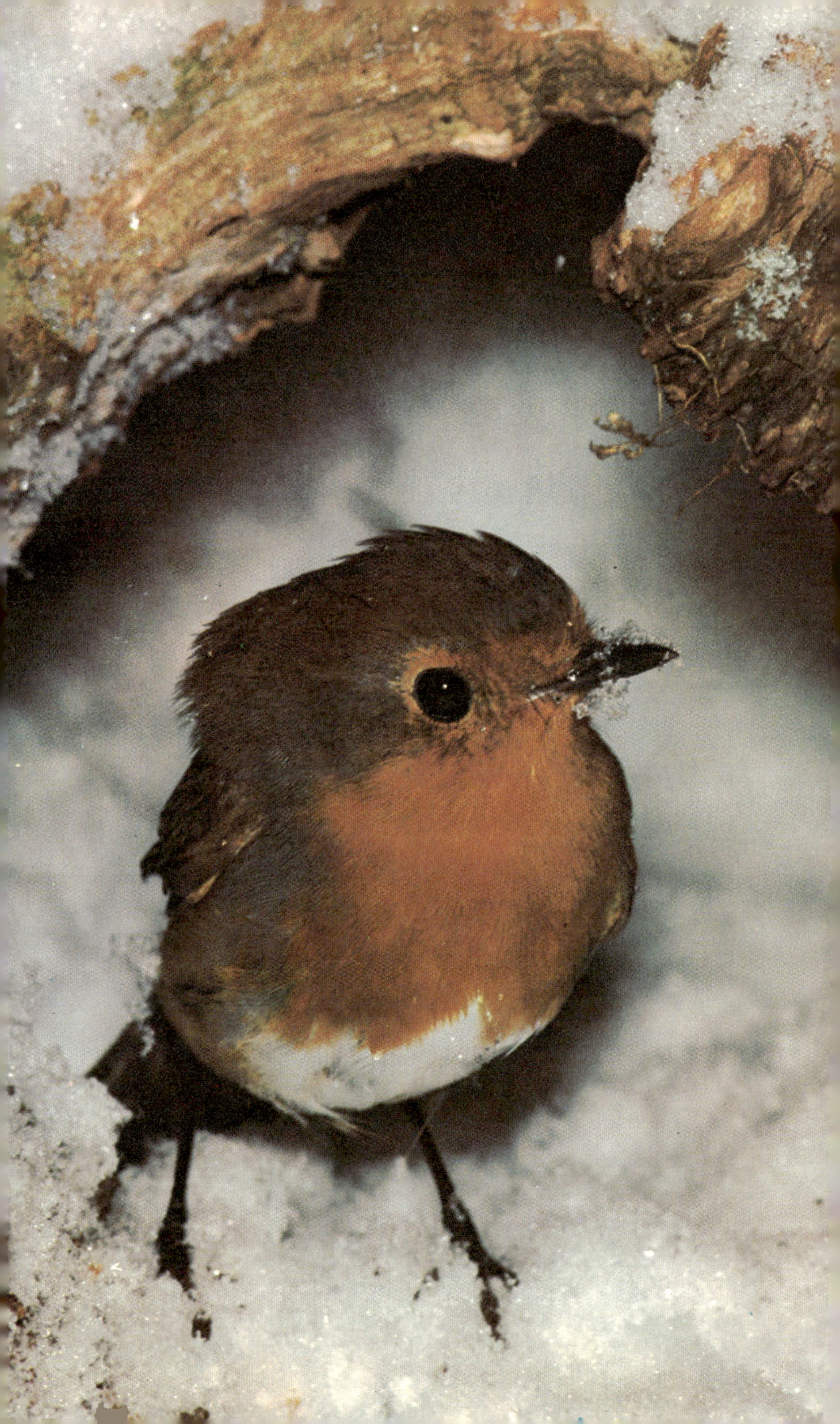

1

2

3

4

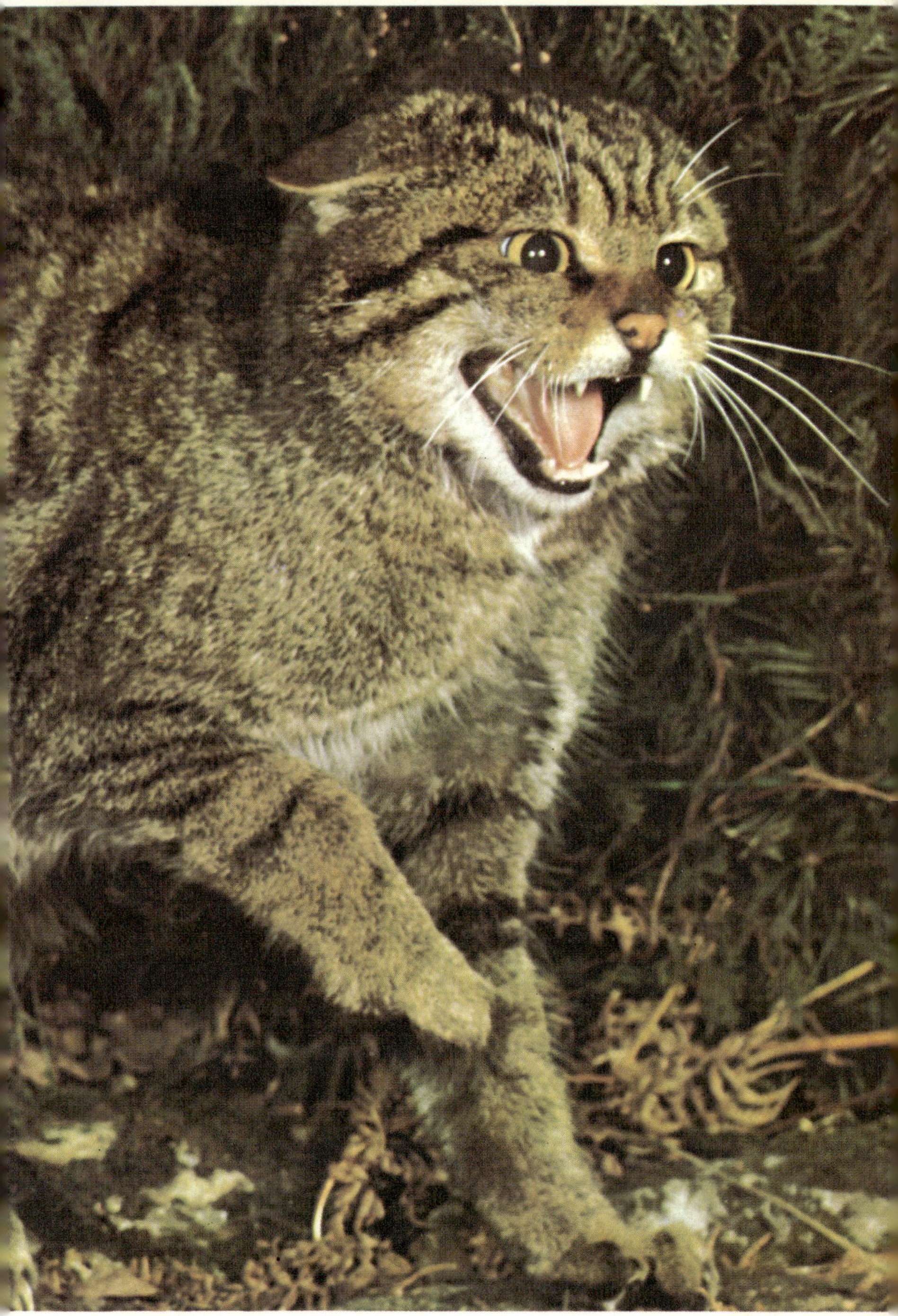

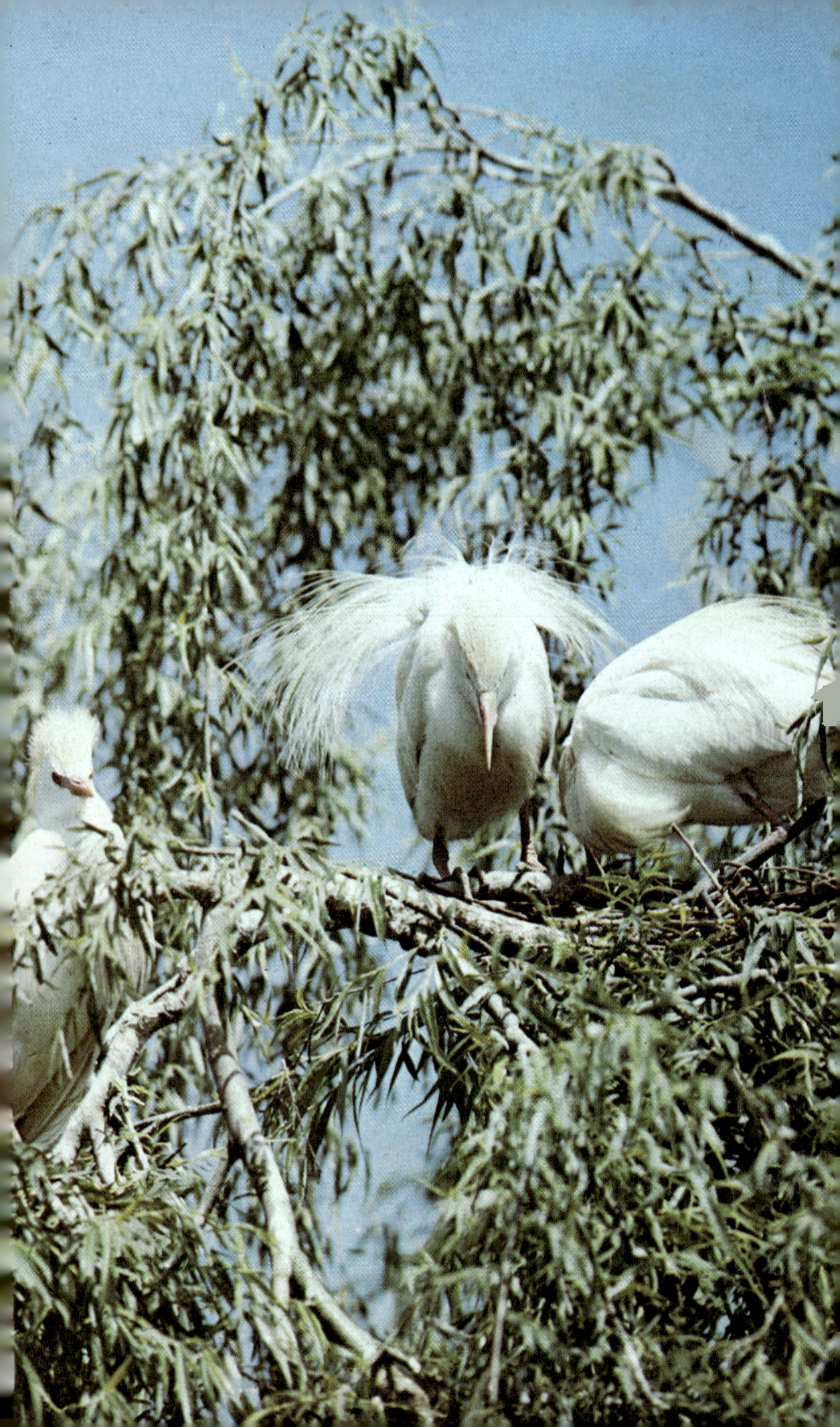

purple and white. Both the shape of the arms and the colour pattern on them have the effect of releasing attack.

The colour blue is the signal which provokes certain members of the lizard family, such as the American fence lizard. These lizards usually hide themselves amongst their surroundings, but during the spring – the breeding season – the males display at each other, bobbing up and down to expose their blue throat and chest to the opponent. This colour display is one of the important rituals of fighting.

Conflict and courtship

In many species the male can be distinguished from the female by some slight colour difference. These differences prevent the male thinking his mate is a rival. The male American flicker, a woodpecker, has a small black patch at the corner of its mouth rather like a moustache. The female has no such patch. If the female of a pair of flickers is captured and given an artificial moustache she is immediately attacked by her own mate, who takes her to be an intruding male and a potential rival. The black patch sparks off his threat display. After the female has had the patch removed, the male accepts her once again.

Slight colour differences also distinguish male and female budgerigars. The cere, a patch of wax-like skin around the nostrils, is blue on a male bird and brown on a female. A female with her cere painted blue is attacked by the male, this small patch of colour being enough to stimulate threat display.

In other species, the male and female do not differ in colouring. The plumage of the male and female wren, for example, is very similar. In species such as these, the female must make special behaviour movements to suppress the male's aggressiveness towards her. Sometimes this means adopting a posture which makes her 'masculine' features as inconspicuous as possible. The female herring gull takes up a begging position characteristic of the young, to divert the male's aggression. This is called a gesture of appeasement or submission. The female bittering uses another method of avoiding her mate's attacks: she swims under him so that in the end he stops trying to attack and begins to court her.

The male south European emerald lizard recognises a female by her smell. The male's aggressive behaviour is sparked off by the gorgeous colours of rivals, especially the ultramarine throat and the green body. A female coloured in these beautiful shades and put in an enclosure with a male is immediately threatened, the male opening wide his jaws with the intention of biting. Once he has smelt the female's distinctive odour, however, he stops

Only male American flickers have a 'moustache', the dark patch at the side of the beak. This black mark acts as a releaser for male aggression and because the female is not distinguished in this way she is safe from being attacked by her mate.

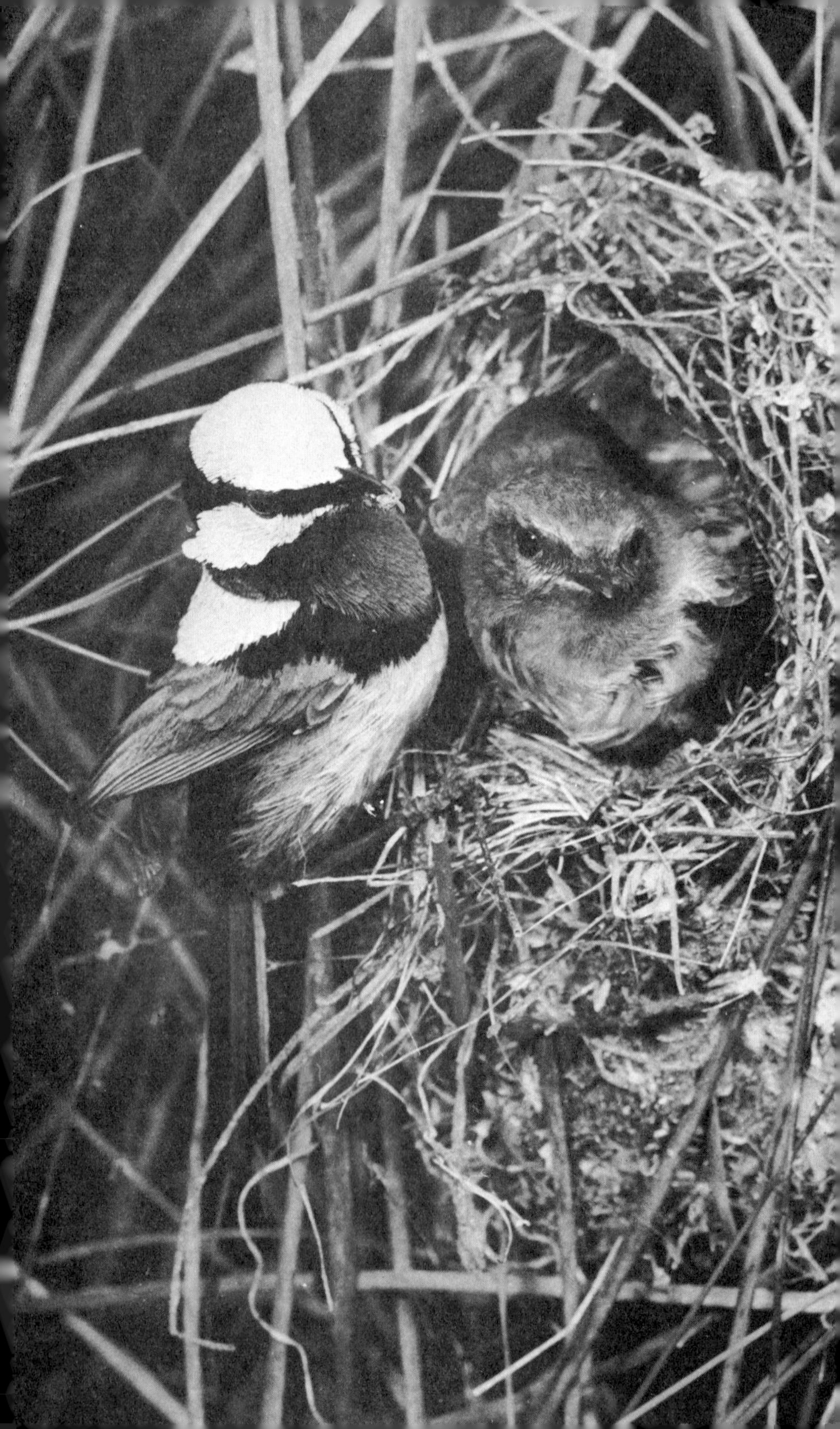

his attack so abruptly that he is quite likely to turn a somersault over her in his confusion.

Although the female must make submissive postures to prevent the attacks of her mate, the male will often tolerate attacks from the female. He does not adopt the submissive rôle but reacts with a sexual self-display and even tenderness. This often happens with bullfinches. It does not mean the female is superior to the male, quite the contrary, for the male's passive behaviour and the way in which he accepts his mate's attacks without becoming aggressive have a very impressive effect on her.

At breeding time, attack, escape and sex are the three driving forces, and all can come into conflict. This is why displacement activities and other such behaviour can be traced in courtship displays. When a female approaches a male, she not only provokes a sexual response but also, unintentionally, she provokes him aggressively. He may well think she is an intruder, and until she makes some submissive posture he does not accept her as a possible mate but remains in a state of indecision, not knowing whether to escape, fight or court the intruder. Some cichlid fish are able to keep their domestic life peaceful while still defending a territory, because the male can combine sex with aggression and the female can combine sex with escape.

Marital tiffs and tussles

During the breeding season three driving forces come into conflict. These are fear, fight and courtship.
Left. *To prevent her mate from attacking and treating her as an intruder, a female blue wren flutters her wings and behaves like a chick. The male responds to this act of appeasement by feeding her.*
Below. *A lion and lioness in fierce disagreement show that most animal relationships have periods of trauma.*

Peace by submission

Submissive postures which divert aggression are important not only in courtship but also in peck-order fights and in the relationship between parents and young. Fights between shrews end suddenly when one animal rolls over on his back and offers his most defenceless parts to the victor. The other shrew then loses interest and moves away. Subordinate baboons will cringe before a dominant male so that his aggressive mood is diverted.

The relationship between a mother and her newly born young is a very precarious one. She must be on her guard all the time against predators or rivals who may enter her territory. But at the same time she must not be aggressive towards her defenceless young. In a night heron colony there are continuous jealous disputes and yet the young fledged birds manage to remain in the colony. The young birds achieve this by begging. Before an older bird can peck, the young one makes begging calls, flaps its wings and tries to seize the older bird's beak, pulling it down for regurgitated food. When an adult bird is not in the mood for feeding its young it flees from its own children, and this is also what a strange rival bird does before the begging young.

As soon as the grown-up young of the herring gull begins to look like an adult bird it starts to annoy its parents. Its very shape stimulates the parents aggressively. The young bird manages to prevent the parents from attacking by showing infantile behaviour, adopting a submissive posture that is practically the opposite of an aggressive posture. The young bird withdraws its neck and takes up a horizontal attitude with its bill pointed slightly upward. This posture becomes less and less effective against the parents as the bird matures but it usually happens that by the time the parental drive has completely waned the young bird is quite capable of looking after itself and does not have to be submissive for attention.

The reverse situation, the submissive posture used by a parent towards its young, may well be just as necessary because the young must be able to distinguish between predators and parents. This kind of submissiveness has been observed in the night heron when the male comes to the nest. Standing on the edge of the nest he makes an exaggerated bow to his mate and young. This shows off his beautiful bluish black cap and at the same time raises the three white plumes on his head, which when at rest are folded down. After this elaborate introduction the night heron is cordially received by the occupants and is allowed to step down into the nest. This appeasement ceremony suppresses the youngs' defence reactions. Once again communication is achieved by means of social signals whose meaning is precise and readily understood by other members of the species.

Normally the three white plumes on the night heron's head lie smoothly over the rest of the feathers (above). *During the rearing of the young, however, they become important signalling devices. Sitting on the nest and searching for food are taken in turns by the adults, and when the changeover from one duty to another occurs at the nest the returning bird raises the white plumes and lowers the head* (below) *in a gesture of appeasement.*

The hunter and the hunted

We have little animals here,
   slow-stepping cousins of stoat and weasel,
Striped skunks, that can spit from under their tails
An odor so vile and stifling that neither wolf nor wild-cat
   dares to come near them; they walk in confidence
Solely armed with this loathsome poison-gas.

from Skunks by Robinson Jeffers

Every animal is in constant danger of falling victim to a predator, whether it is the deer that is so silently stalked by the tiger or the male spider that is eaten by its mate. However, the hunted fight back. The skunk for instance is quite capable of discharging a stinking secretion at the face of its enemy.

Wasp eats spider, fish eats insect, snake eats frog, owl eats vole, lion eats zebra and so it goes on and on. Constant battle is waged between the hunter and the hunted. This kind of war is not vicious, nor spiteful nor blood-thirsty. An animal must eat to stay alive, and if this means killing another creature then this will be done as a matter of course. Emotions play no part and an animal only kills as much as it needs to satisfy its hunger and never for the sake of destroying life.

It is surprising what a useful function this eater–eaten relationship serves for both predator and prey species. The hunter tends to remove the surplus population of a species so keeping a steady level of numbers. Because the hunter usually removes the old or infirm creatures which are less able to escape, this is likely in the end to strengthen rather than weaken the health of the hunted species as a whole. In territorial animals it is those that do not have a permanent home or territory that tend to be killed, and even if they are not killed in this way they are not likely to reproduce unless they can find a home. So all in all the war between predator and prey is of use to both populations, and represents a delicately-balanced system in the economy of nature.

Very few animals are confined to eating only one kind of prey. If one source of food is scarce then the hunter will eat something else. The myxomatosis plague of rabbits in Britain did not mean that the foxes starved to death, they quite naturally turned to their other sources of food of rats and mice and even snails and beetles.

On land and in the sea and fresh water there exist food chains – the intricate relationships between animals feeding upon each other and on plants – in which small animals tend to eat small prey, large animals large prey. A small spider will eat a small insect, a large spider a large insect. Each animal feeds on certain types of food best suited to its needs. A lion does not bother to hunt mice; he would need too much energy to catch sufficient to satisfy his hunger. Man is one of the few creatures able to deal with food of all sizes. Like ants and wolves he can team up with others of his kind if necessary to hunt prey larger than himself.

Target-shooting fish

Man of course is a proficient hunter, because he is aided by all kinds of weapons and devices of his own invention. There are certain animals which also impress us with their hunting ability, and one of these is the archerfish. This small fish lives in the mangrove swamps of river mouths and small bays in southeast Asia and northern Australia. It catches its food of flies and other insects by shooting them

Precision shooting: the archerfish of Indo-Australian waters is famed for the accurate way in which it is able to capture insects resting near the water with a jet of liquid shot from the mouth.

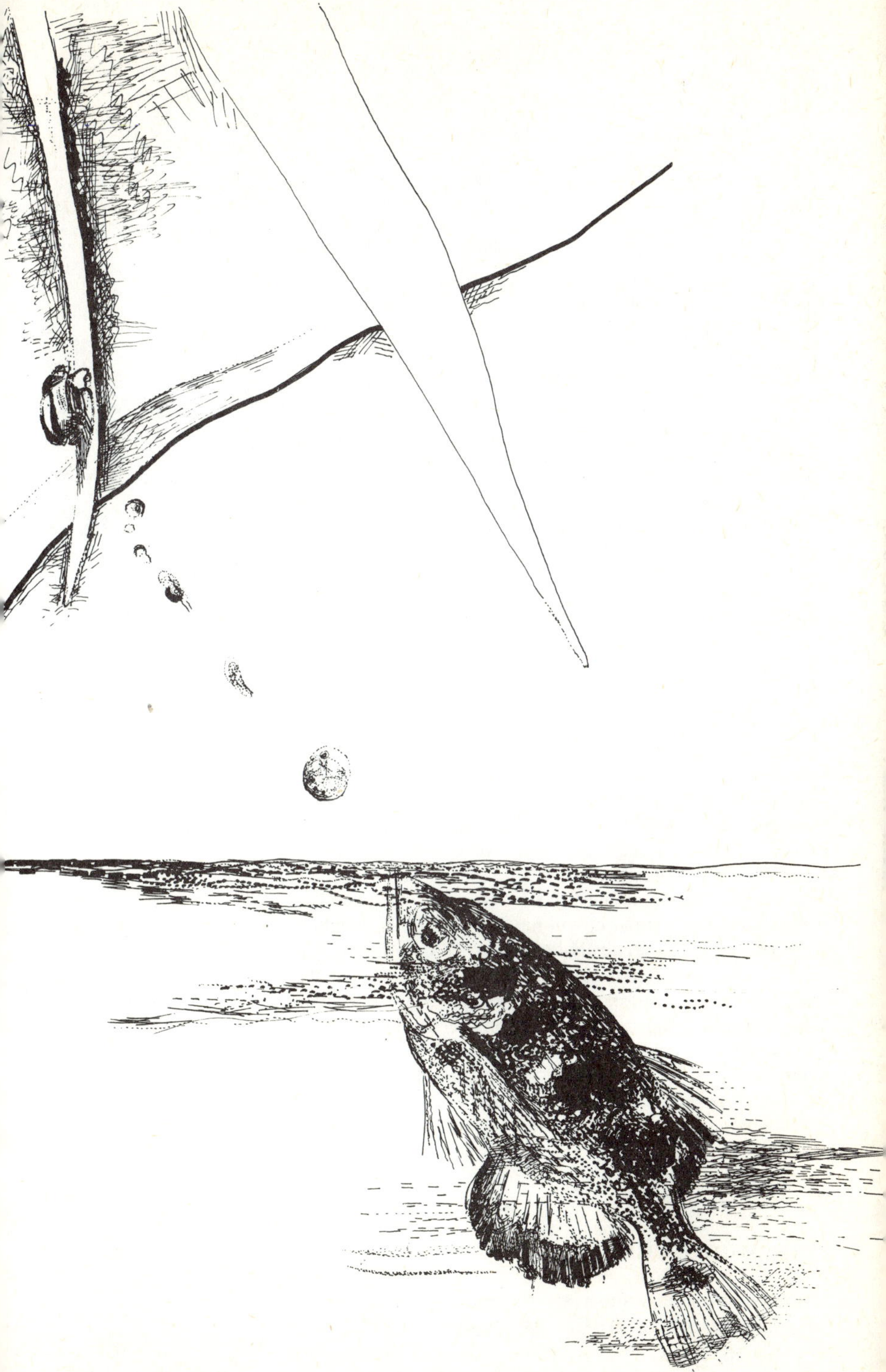

The lion drags his kill to cover. He is yet again the victor in the age-old war between hunter and hunted.

down. To do this the archerfish swims just below the water surface on the lookout for insects that may be resting or sunning themselves on a water plant. When within shooting range the fish positions itself to take aim. It cocks the whole body like a gun-barrel, all the time taking care not to penetrate the water surface and so scare the insect. The 'shot' is a fine jet of water that is fired fanwise so that it covers a large target area. In this way the fish is most likely to have a direct hit almost every time. Should the first shot miss the fish can fire up to seven times in quick succession. But the archerfish is not likely to be given this opportunity as the insect will have flown off. The archerfish prefers to fire vertically because when it looks straight up out of the water it can see the exact position of the target. This is just the same as when a stick is dipped straight into water and viewed from vertically above. It does not look bent. If the stick is dipped in at an angle, however, it appears bent due to the refracting of light rays at the water surface. Aiming vertically also has another advantage: when the insect is hit it drops straight into the fish's open mouth.

Carnivore tactics

Other impressive hunters are the group of flesh-eating mammals known as the Carnivores (not to be confused

With determined tread, a leopard starts to run towards the prey which it has been so silently stalking.

with the word carnivore used simply to describe any flesh-eating animal), and the birds of prey. The Carnivores, such as wolves and lions, use their powerful teeth and claws to overcome prey, while birds of prey such as eagles, have the advantage of swooping flight to surprise their quarry and are armed with a sharp beak and strong grasping talons. Some invertebrates can also be ruthless hunters, like the army ants described in the first chapter and many spiders.

Every animal has its own individual tactics and strategies for catching prey, whether by tirelessly chasing the prey, by hiding and pouncing on the unsuspecting victim, by co-operative attack or by setting traps. The lion, tiger, cheetah and other cats are animals that rely on their cunning and stealth for obtaining their food. The tiger usually hunts on its own moving silently through the night in search of deer, cattle, horses and many other domestic animals. It can keep up a steady speed quite effortlessly catching large prey that is on the move. Its smaller victims it takes by surprise, stealing up on them and then bounding rapidly over the last few yards to make a kill.

The cheetah is the cat renowned for its sprinting ability. It too is a solitary hunter but catches its food during the daytime for it needs to be able to see its prey of gazelles from a distance. It crouches on the ground inching up towards a group of these unsuspecting antelopes and then hurls itself at full speed towards them. Gazelles are agile creatures and may escape the cheetah. If the cheetah is quick enough, however, it will catch up with them and bowl one over with a blow from its paw.

Most cats hunt on their own; the lion is the exception for often family groups or prides will join in a well organised hunt. An experienced male moves upwind of a herd of antelope and growls, giving the uneasy animals his scent so that the herd stampedes towards the rest of the hunting party that lurk among the tall grasses. It is usually the females that make the kill in group hunting, and they learn to deliver the 'death-bite' quickly. Turns at feeding always start with the dominant male, then the females and finally the young. A good kill, such as a zebra or gnu, could satisfy a family of lions for several days. Lions have a healthy respect for large prey such as gnu as they are vulnerable to butting heads and jabbing horns. In such a case they prefer to attack from the rear, clawing at the legs and disabling the prey before making the kill.

Wolves hunt in family parties or packs. They do not depend on stealth but on their enduring power to run down their quarry. The wolf is a tireless hunter, ruthless and ferocious and endowed with great courage, fighting ability and intelligence. A pack may consist of up to twenty-four animals and is then capable of hunting down large animals such as North American moose or elk. The success of a communal hunt depends on a high order of

Too close for comfort

A single husky is no match for a polar bear as this dog has just realised.

co-operation between all the pack members. Wolves will run in single file for miles until they sight a lone moose and then they spread out several yards downwind, standing still like pointers. Suddenly they all rush together, touching noses and wagging tails. The large leader wolf heads straight for the quarry while the rest fan out behind. If the moose tries to escape they close in tearing at its rump and flanks. The moose kicks desperately and tries to run again but is finally cornered and is killed within minutes. The wolf eats enormous amounts and may sometimes hunt on its own. It will often run after a moving herd of antelope, patiently waiting for an old or sick animal to get left behind before moving in to attack.

Like the wolf the Cape hunting dog or African wild dog is a persistent chaser. The dogs live and hunt in packs but break a basic rule of nature in that they seem to kill just for the sake of it. They excite our disgust because they eat their prey even before it is dead. They chase stubbornly after their quarry, a gnu for instance, and when one dog is exhausted others will take over, pressing hard on the heels of the unfortunate hunted animal, until it is worn down and weak from loss of blood from where the dogs have snapped hunks of flesh out of its flanks.

Highly dramatic paintings resulted from the white man's first encounters with tigers and probably led to the false belief that all tigers were likely to attack without provocation.

Man-eaters

There is yet one other animal that to most men and women has even more distasteful and horrid eating habits than the Cape hunting dog. This is the piranha fish that lives in rivers of South and Central America. Piranhas hunt in shoals sometimes of several hundred. They usually feed on small fishes but the most gruesome stories arise from their attacks on animals entering or accidentally falling into the water where they are swimming. Stories are told of a cow or pig being stripped to a skeleton in only a few minutes. And naturally these fish do not stop at cows but will just as happily take chunks of flesh out of a man. These stories may be wildly exaggerated for in some villages the children play quite happily in piranha-inhabited waters, and women do their washing in the rivers. In other villages, however, men have scarred legs where a piranha has quite cleanly bitten off a piece of flesh. The fishes seem to vary in their ferocity from river to river. The jaws of a piranha are so strong and the teeth so pointed and sharp that they can chop out a piece of flesh as neatly as a razor.

The fishes most feared by man are without doubt the sharks. It is usually through the diver's own carelessness or the bather's lack of watchfulness that life may be lost. The point at which a curious shark may become an attacking shark is when a diver is carrying a recently speared fish that is leaving a blood trail. The shark is usually interested only in the fish and not the diver, but even a normally docile species can have its temperament radically affected just by the presence of blood or fish juices in the water. Sharks have also been known to attack when hampered in some way. Examples are known where a diver has taken hold of a shark's tail. The shark can hardly be blamed for looking after itself.

Most shark attacks occur in warm water in tropical areas, or in temperate zones in the warmer months. Records of sixty-one Australian attacks showed that slightly more than half took place at the warmest part of the day: between two and six o'clock in the afternoon. One surprising outcome of this study was that the sharks seemed to prefer men to women in a ratio of twenty to one. With both results of the survey, however, one must consider the circumstances. More people bathe in the warmer hours, and men tend to venture further out than women. More study is needed to prevent any more loss of life.

The greatest man-killer of all is the tiger. The majority of tigers that turn to killing men do so because circumstances make it impossible or more difficult for them to catch and eat their normal food. In most cases this happens because the tiger is wounded in some way so that its normal agility is impaired. A tiger may, through old age, turn into a man-eater, but this is rare. One tiger that killed twenty-four people before being killed itself was found to have several porcupine quills in one of its paws. These were probably a great handicap when the tiger tried to get its normal food. Once a tiger has tasted human flesh it may well realise what easy prey man is, and the animal quite often remains a man-killer for the rest of its life. A tiger loses all fear of man after becoming a man-killer and takes to hunting during the daytime when most humans are moving about freely. Leopards that become man-eaters, however, never lose their fear of man and continue to kill at night, and for this reason they are much more difficult to track down.

Man interferes

Man himself can be a very ruthless hunter. He has always taken a certain proportion of animals for food and other purposes, and until the last two hundred years this has not endangered species in any way. Man has been in exactly the same position as any other animal predator. But once modern weapons, especially guns and powered harpoons, came into regular use, the balance was upset and excessive killing took place. The extinctions of over-hunted species begun in the nineteenth century have continued to the present day and are one of the scandals of our age.

It may be that the unseen effects of man's interference with the balance of nature are even more far-reaching.

A trumpet shell or triton devours a crown of thorns starfish which is consuming large quantities of the coral. One reason for the massive increase in starfish population and the subsequent loss of coral is that the triton, the only animal that preys upon the adult starfish, is collected in great numbers by tourists.

An alarming problem developing on some of the world's coral reefs may well be connected with human activities. Scientists are desperately trying to find some way of dealing with the disastrously high number of crown of thorns starfish in the Pacific.

In 1969 biologists became aware of the fact that this echinoderm had undergone a population explosion and was rapidly eating its way through the coral reefs of the Pacific. But from about 1962 onwards reports were continually coming in of the abnormally high population of the starfish. In 1966 5,750 were counted in a hundred minutes on one small section of a coral reef, and one diver caught 27,000 in fifteen months.

This large sixteen-armed spiny starfish has already devoured its way through a quarter of the Great Barrier Reef. Its dietary statistics are quite horrific. It eats twice its own area in a night and destroys a square yard of coral in a month. This is just one starfish; imagine thousands of them voraciously satisfying their hunger on coral polyps. It will take hundreds of years before these beautiful corals will be replaced, even if they get the opportunity to grow again. This is not all. The corals are important not only to the fishes that make their homes amongst them, but also to the inhabitants of the nearby islands. The people rely on these fishes as a source of food, and on the corals themselves for the tourists they attract – an important source of income. All this will end when the reefs are just a mass of lifeless skeletons.

It is very noticeable that this starfish is most numerous wherever man is present. The cause may be man-made pollutants pumped into the sea. These pollutants may be slowly eliminating the planktonic predators of the starfish larvae which occur in vast numbers during the spawning season. Or the population explosion may be a result of the over-collection of the triton shellfish, the mollusc that preys upon the adult starfish. The third alternative is that this explosion is due to a basic change in the life history of the starfish, in which case control is not possible and the coral of the Pacific will slowly become extinct. The crown of thorns is a most efficient predator. Its methods can be compared to the bulldozing technique of destroying all in its path.

Insect cannibals

Insects and their relatives may produce thousands or even millions of young. If all survived, they would soon cover the earth several deep. The only reason the insects do not take over the world is that other insects kill them first: 'the best insecticides are other insects.' Ladybirds and their grubs eat aphides, robber flies catch bees, ichneumon flies parasitise the caterpillars of white butterflies, ant-lions trap ants – the list is endless. In their miniature world the insects carry on continual war with each other.

The life of insects and spiders also shows us some of the rare examples of animal cannibalism. They happen in certain species when the male gets himself mistaken for prey by the female at mating time. Aggression and courtship often go hand in hand, and the male must act in accordance with strict rules if he is not to be eaten after fertilisation.

In most species of spider the female is bigger than the male, but the belief that she always eats her mate is quite unfounded. The male is polygamous, that is he has more than one mate in the reproductive season, a feat that would be impossible if the female ate him immediately after mating. But having fertilised many females a male spider will become enfeebled and moribund and this is when he is devoured.

The empid flies flourish in the cooler climates all over the world. The males are particularly cunning in their approach towards a female. Many of them present her

Having carefully parcelled him up in thread a female garden spider begins to eat her mate. Male spiders are polygamous, that is they mate with more than one female in the reproductive season. After several matings the male becomes weak and it is then the female gets the chance of an easy meal.

Opposite. *An angry tiger is an awesome foe. Although man-eaters are rare, once a tiger has tasted human flesh it loses fear of man and becomes a dangerous killer.*
Overleaf. *Intelligence and co-operation are needed for animals to hunt in groups. In winter when the snow is deep wolves move about in packs chasing after prey such as moose. The wolves surround the animal and harry it until it panics and bolts. The wolves then move in and tear at its unprotected flanks.*
Page 100. *The mantis is a voracious feeder, victims may be moths like this or other mantids – a female has no compunction in biting the head off a male during mating.*

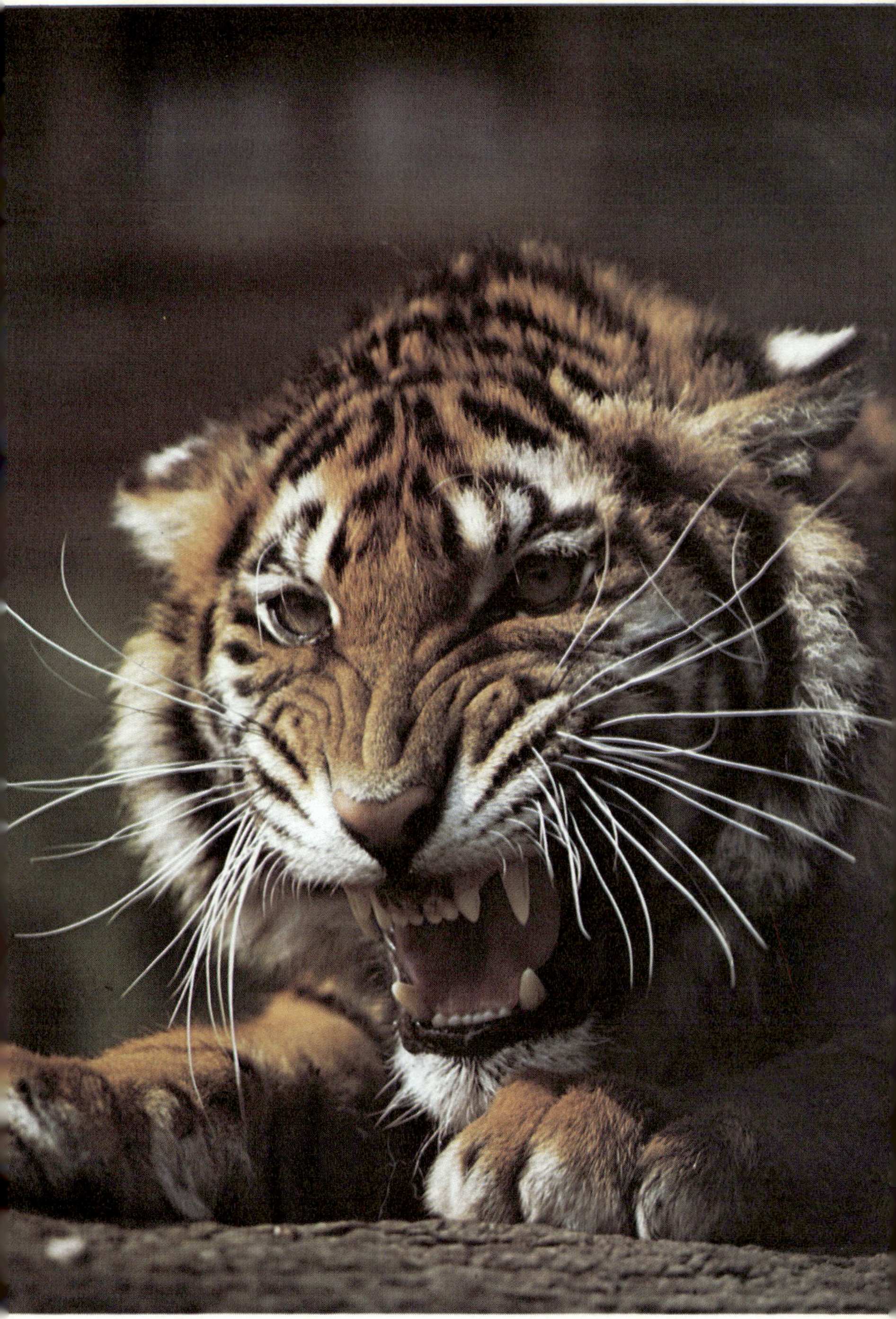

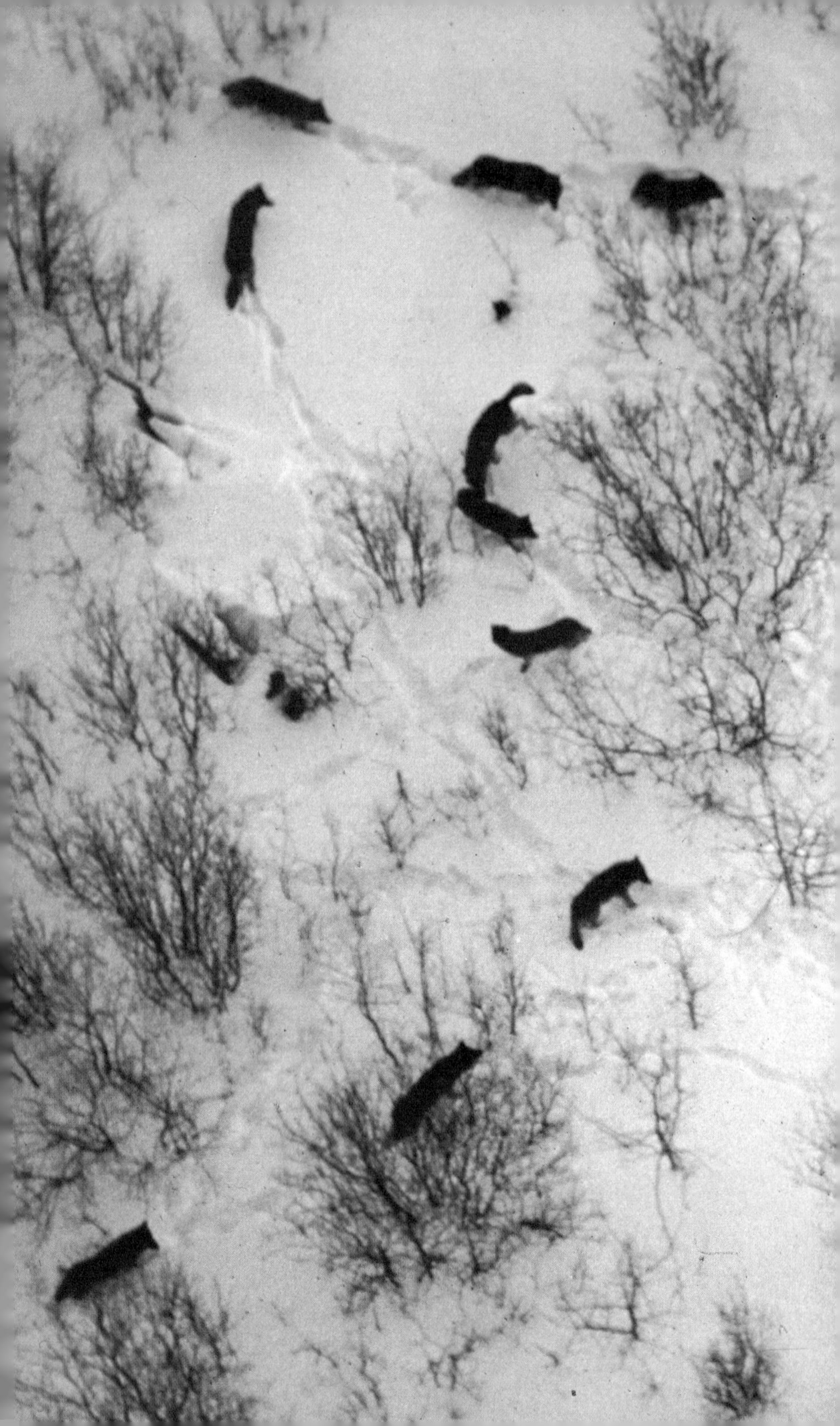

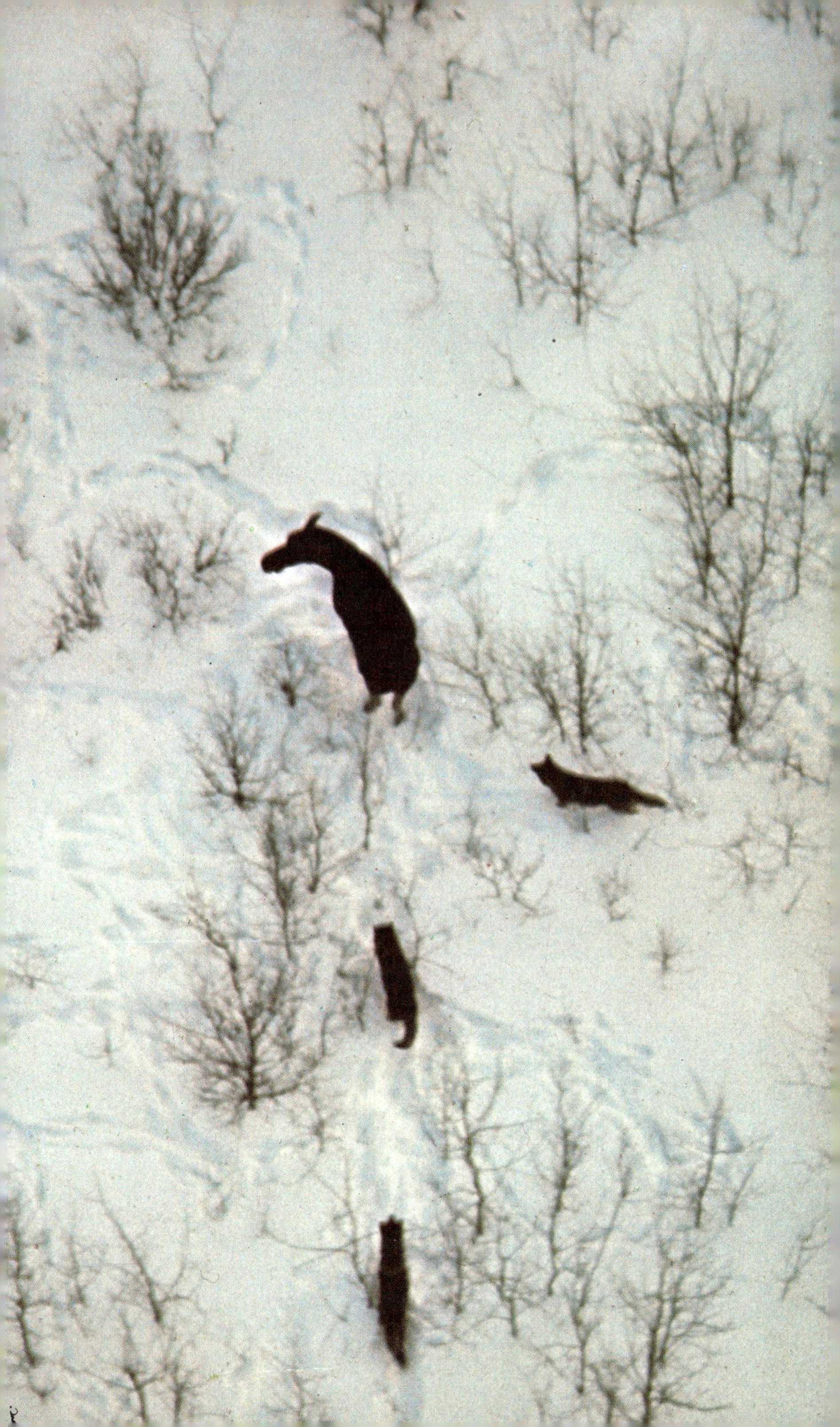

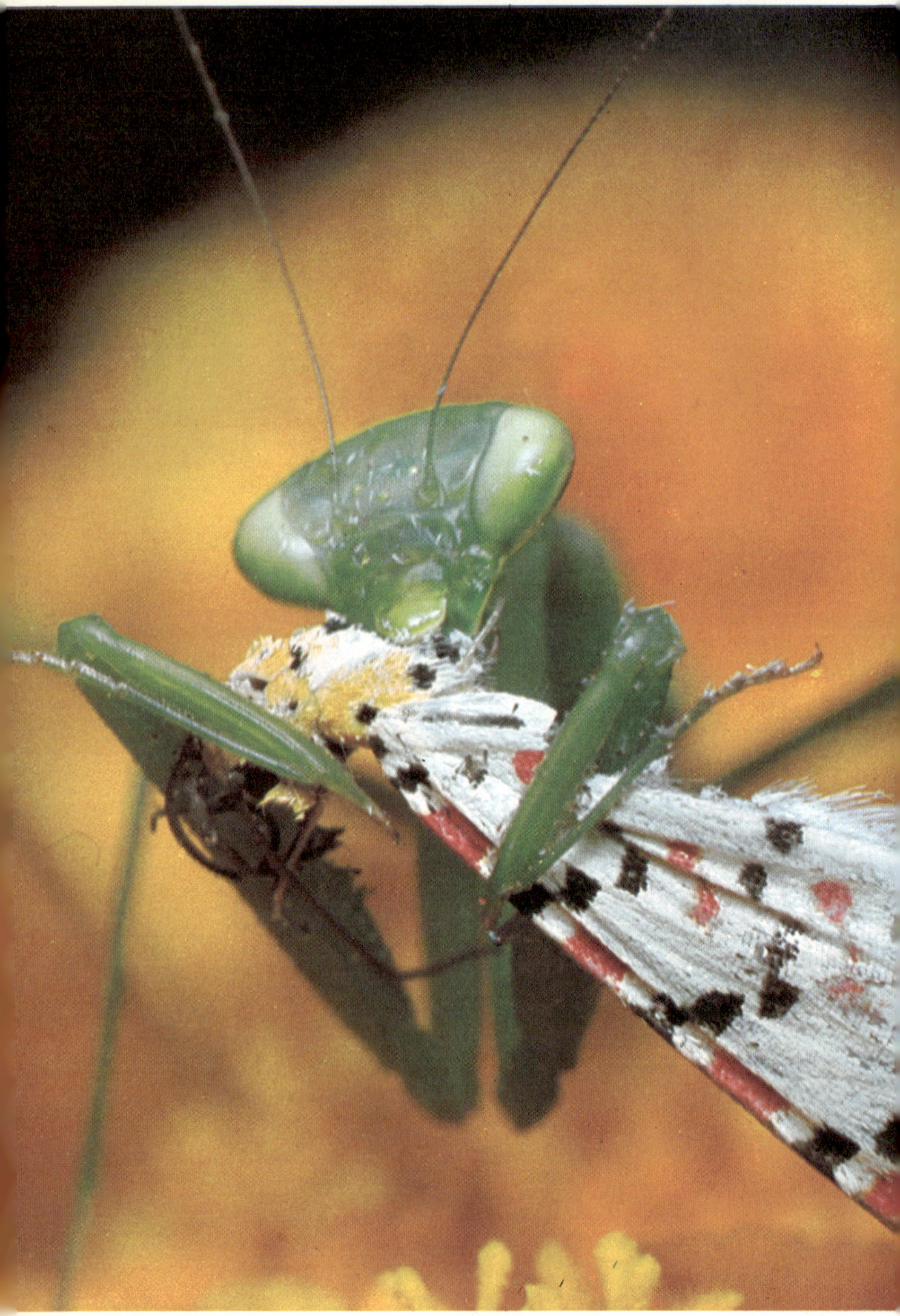

with a slaughtered insect which she eats while he mates without fear of being eaten himself. But most remarkable of all is the way the male tailor fly distracts his mate's attention during fertilisation. He spins a lovely little ball of silk during flight between his middle and hind legs. It is well established that the male offers the silken ball to the female as a part of courtship. The sequence and significance of events have been much debated. Some scientists believe that the female is mesmerised by the optical stimulus of the silk and the male can mate with her unharmed.

The female mantis, however, quite justifies her cannibalistic image of mate-eating. Should the male approach her directly from the front he will almost surely die. He will have his head bitten off almost immediately. His best tactic is to inch up to the female from behind, moving so cautiously that it takes an hour to go a few inches. As soon as he is near enough he makes a short hop and clasps the female. All goes well unless the pair are disturbed. If this happens she will start to eat him, first biting off his head. Mantis copulation is controlled by a nerve centre in the male's head which inhibits mating until the female is clasped. When the female decapitates him, the nerve centre control is removed so he loses all his inhibitions and his body continues to copulate automatically. The

A false step usually proves fatal in mantis courtship. Here a female mantis eats the foreparts of a male unwise enough to approach her from the front; more cautious suitors survive by advancing towards the female from behind her abdomen.

female has much to gain by attacking her suitor. She not only ensures fertilisation of her eggs but also nourishment for her developing ovaries: the brave male loses all.

Wasp against spider

Methods of attack are only half the story in the battle between predator and prey. Defence is equally important. Animals constantly devising new methods of attack and defence against each other are the hunting wasps and the large wolf and trapdoor spiders.

At first glance the odds seem very much against the smaller wasp, but whatever method of defence the spiders use the wasp always seems to penetrate it. The female wasp out hunting for food for her offspring has only her sting as defence against the eight clutching legs and powerful poison fangs of a spider. Nevertheless the wasp is usually the victor because she is able to immobilise the spider with a well-aimed sting in the spider's nerve centre. This counter-attack puts the spider's legs and jaw-pincers out of action and the wasp can finish off her prey without more ado.

This kind of war between wasp and spider has gone on for millions of years, with new weapons and counter-weapons evolving in each species to combat the other. One very ingenious method of defence of a cave-living trapdoor spider is to spin radiating alarm lines or trip wires from the trapdoor of its tunnel home. An approaching wasp cannot fail to touch them and the split second delay is enough to allow the spider to take evasive action by retreating to its home and shutting tight the trapdoor.

The most defenceless animals are the young. Birds spend a great deal of their time defending their chicks. Should a dog or a human come too near a herring gull's nest the mother bird will 'charge' the potential enemy,

Right. *A marsh harrier is mobbed by two common terns. Many birds use this method of warning off an enemy.*

Below. *The reaction of musk-oxen to danger is to form a protective phalanx, known as the 'hedgehog formation' with the cows and calves shielded by the big males. Wolves and other predators are usually deterred by this, but men with guns have been quick to take advantage of the oxen's reluctance to flee, and there has been much indiscriminate shooting of them.*

swooping down over the intruder's head again and again. Usually the bird does not touch the intruder but will give the impression of preparation for real attack by lowering one or both of its feet a fraction of a second before flashing past the enemy.

Many birds have special alarm calls which warn the young of approaching danger. For instance jackdaws 'rattle', making a distinctive grating sound that is sharp, metallic and echoing. The alarm signal need not necessarily be a sound but may be of a chemical nature such as that found in many social fish. When a pike or a perch snatches a minnow from a school the other minnows scatter immediately and remain on the alert for a long time. The minnows are warned of the danger of a predator by smelling a substance that is released from the skin of the slain fish. The substance is only found in minnows and so only minnows react to it.

Deterring the enemy

Animals co-operate not only to hunt down prey but also to ward off an enemy. Communal attacks against a predator are best known in birds. Jackdaws, terns and many songbirds will mob an enemy flying through their territory. They will gather round a sparrowhawk, little owl or prowling cat, or fly above the predator in a dense cluster trying to keep well above it before swooping down. Unless the hunter is desperate for food it will soon give up and go away as the mobbing attacks get more forceful. If the hunter is really hungry then these attacks do not disturb it too much but its attention becomes distracted from detecting other prey. Mobbing can also be effectively used by larger animals. Even zebras will molest a leopard caught where cover is sparse if there are enough zebras to join in an attack. Musk-oxen form a defensive circle or phalanx if attacked by wolves, with the calves and cows in the centre. Musk-oxen *en bloc* in this way are enough to deter the most hungry of predators. Individual wolves are gored or trampled or a pack of wolves will be driven off by a combined charge.

Other ways of defending against an enemy are to use weapons such as teeth, claws, horns, antlers, armour or quills. More complex devices include secreting foul-smelling substances, or shedding part of the body such as the tail of a lizard or the arms of a starfish. Camouflaging the body to blend with the surroundings gives passive protection, whereas very conspicuous body colours warn attackers to keep away. Some animals freeze into a rigid position or sham death when about to be attacked. All these methods of defence have proved quite successful in animal warfare.

Skinplates and hairspines

Armadillos and pangolins are covered with an armour of thick plates. These 'plates' are formed of very hard, tough skin, so tough that they protect the animal from the teeth of jackals, lions and other members of the dog and cat families. The armadillo and pangolin are able to roll themselves up into a ball so that their soft parts are protected by the hard armour. The most persistent predator usually tires of trying to 'crack open' the body. Hedgehogs and porcupines have spines and we all know how difficult it is to try and pick up a rolled up hedgehog.

Like a jigsaw puzzle, the parts of the armadillo's tough armour dovetail into each other when it curls up, making the animal almost invulnerable to would-be attackers of all kinds.

But the porcupine is the most impregnable. It has an array of quills on its back which can be raised by powerful muscles when the animal is disturbed. If the enemy does not retreat the porcupine charges backwards at its foe so that the quills, held out like a battery of lances, make contact, become detached and remain embedded in the predator. This can be quite disastrous for the predator. A tiger was killed by porcupine quills which perforated its liver and lungs in such an attack. Even quills embedded in an attacker's face or leg can prove fatal as the animal finds it more difficult to find food, and may die through lack of food and exhaustion.

Teeth, claws, horns and antlers are used as weapons in many territorial fights. In these battles they must be used with reserve but in the hunter–hunted fights where an animal's life depends on its ability to defend itself then these weapons are used with full force.

The self-amputation of the tail is a well known reaction to attack in lizards but several small rodents, dormice and field mice, are quite capable of shedding part of the tail in emergency. Unlike the lizard which actually drops its tail with a clean break a mouse only sheds part of the tissue in which the tail is enclosed. The predator is left with the outer covering while the mouse and inner part of the tail escape. This denuded portion eventually withers away in a few days and drops off. The lizard can at least grow part of a new tail, the mouse cannot. The lizard's sacrifice is of great advantage, because the tail continues to writhe for half a minute, distracting the attention of a predator while the owner makes good its escape.

Rabbits and deer can stop dead or 'freeze' in their tracks

The quills of the American porcupine normally lie flat along the body, but can be erected when the animal is aroused. Detached quills may become embedded in an enemy – a useful deterrent.

Whichever way you look at it, the skunk's defence mechanism is an efficient one. The striped skunk (top) plays no fancy tricks but simply directs the anal glands which produce the stinking secretion straight at the foe. Spotted skunks (below) *perform a handstand for maximum effect.*

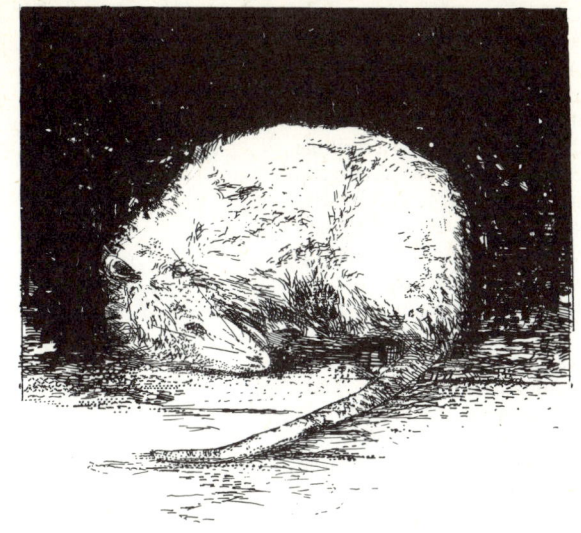

Playing dead may not seem the surest way of escaping enemies, but it seems to work for the Virginia opossum. For some reason, predators are put off by the 'lifeless' body which seconds before was a living, breathing, potential dinner, and they lose interest. The danger past, the opossum walks off.

at the sight of an enemy and in this way perhaps escape detection. The American opossum can go one step further and feign death or 'play possum'. If attacked or frightened the opossum first hisses, growls and bites, but if this does not work it falls down on its side in such a position that it really does look dead. Its withered-looking ears and bare tail add to the effect and as a final touch the animal draws back its lips to expose its teeth in a horrible grimace. The hunter, after a few sniffs, usually moves on. Some minutes later the 'corpse' comes to life again. It is not clear why a hunter should lose interest in the opossum simply because it has stopped moving, but somehow the trick must have proved of advantage to the species.

Animal stinkbombs and chemical guns

Chemical warfare was not invented by man. There are many animals, particularly insects, which deter enemies by chemical means. They may actually taste bad, which means that the lives of a few individuals must be sacrificed before a predator learns to leave the whole species alone. Or they may squirt out a repellant, like the ants which eject formic acid from the end of the body or the soldier termites which spray a chemical from the modified head.

Taking the prize for efficient chemical defence are the bombardier beetles of Africa and Asia. The rear end of the body is shaped like a gun, which produces a spray of devastating effect when the insect is attacked. Almost all predators, including ants, mantids and spiders, are subjected to seizures when they come into contact with it. The 'gun' can be aimed in any direction by the bending of the last segments of the abdomen, and what is more it can repeat the shots rapidly. Three chemicals stored separately in the abdomen are brought together in the presence of an enzyme and the result is an explosion within the body which is clearly audible, and visible as a mist of repellent.

The most horrible way of deterring a predator is the North American skunk's method of discharging a stinking secretion from its very large anal glands. The skunk can shoot this substance for up to twelve feet and with astonishing accuracy. The stances they use to do so vary with the species. Spotted skunks stand on their front legs and curl their bodies towards the target. Striped skunks simply turn round, presenting their glands in the most direct manner.

The civet and mongoose can also empty stink glands when danger threatens. Many animals have musk glands which they discharge when frightened. The peccary has one on its back a little way in front of its tail. When the animal becomes excited the hairs of its back and neck bristle and the gland emits the musky secretion which can be smelt several feet away.

Not only do skunks have stinking secretions as a means of threatening an enemy but their striking black and white coats also act as a warning signal. Some animals do their utmost to conceal themselves amongst their surroundings to avoid being eaten while others use their brilliant striking colours as a warning that attack will bring unpleasant consequences. The mantis will quiver its body so that it looks like a leaf blowing in the wind but if danger becomes more imminent it will resort to its threat display of suddenly stretching its wings and forelegs and raising its abdomen so that the special bright colours of its wings, normally hidden when the wings are folded, are now in full view. This display is often enough to repel small lizards and birds from attack.

Flight is perhaps the most obvious method of escaping a predator, but unless the hunted can run faster than the hunter then flight is quite obviously useless. This is when the more effective avoiding techniques such as those already described are adopted. However, flight is usually the first reaction to danger. Many animals allow predators to approach to a certain distance before fleeing, but if the attacker manages to get within this distance they may turn and fight, baring the teeth, hissing, or using whatever defence methods they have in their power. A cornered animal tries to look more frightening than it really is, and often uses the same threat postures employed against rivals.

A few animals put up no fight at all, like those hoofed mammals which, pursued by hunting dogs, will stop running as soon as they are overtaken and allow themselves to be eaten alive. Further flight would be of no use and on this occasion at least, they are destined to be the losers in the age-old fight between the hunter and the hunted.

Animals enlisted by man

A gentle knight was pricking on the plaine,
Ycladd in mightie armes and silver shielde

His angry steede did chide his foming bitt,
As much disdayning to the curbe to yield:
Full jolly knight he seemd, and faire did sitt,
As one for knightly giusts and fierce encounters fitt.

from The Faerie Queene by Edmund Spenser

The Persians developed the first cavalry in the ninth century B.C. Since then the horse has been used by man in combat, from the huge armoured beasts of the Crusades to the light cavalry chargers of the last century. Camels, elephants, carrier-pigeons and dogs, too, have played vital rôles in man's warfare.

Man is the great war-maker, but he is not content to wage war alone. Since early times he has enlisted aid from the animal kingdom, often in the most ruthless ways, to help him to victory in his battles. The use of animals in war is as long-standing as the history of warfare itself. An army of men cannot possibly carry all the weapons, spare parts, ammunition, clothing, tents, artillery and food stocks required to keep its human components alive and to wage war efficiently. Man has therefore always used animals as beasts of burden to supply his armies in the field.

But when it comes to the battle itself, man is hampered in another way: by his lack of mobility. Battles are most likely to be won by the side which can marshal its forces where they are most needed as fast as possible. The desire for mobility meant that man quickly learned to utilise the speed of the horse for his battles, and the horse soon became all-important in early skirmishes and fights.

The horse, in fact, must surely defy all comers as top animal in the military league. There have been many changes throughout history in the types of warhorse used in combat. The earliest instances of man riding a horse into battle are lost, but sculptures dating from the ninth century B.C. show Assyrian horsemen armed with bows and arrows riding to war.

It was the Persians however who developed the first real cavalry, their swift, strong horses providing them with the speed for organised manoeuvres and the new and impressive ability to fight on the move. The horse became the winner of battles and cavalry became the most important part of an army. The Persians themselves are reputed to have had eighty thousand fighting horses in 480 B.C., which at that time made up an invincible force.

The mighty warhorse

Once the success of the cavalry had been established, the use of armour for the riders started to develop. Such was the enthusiasm for body armour that heavy plated metal became used in great quantities. This meant that the light, swift horses previously used were not strong enough to carry the additional weight. By about the thirteenth century the armoured knight with his huge lance was the most important part of battle. The chivalrous knights that fought in the Crusades and in other battles relied to a great degree on their horses for personal success. The knights' armour – metal plates and chain mail that almost

Far left. *The armoured knights of the Middle Ages usually came from noble families as it was expensive to provide a good battle horse with all its ornamental trappings.*

Left. *Italian craftsmen made this suit of armour for King Henry VIII in the early sixteenth century. The horse too is protected in much the same way as the rider, with Flemish armour.*

An important part of most nineteenth century armies were the regiments of lancers. These heavy cavalry units were used in tactics of shock against the enemy.

completely covered their bodies – was incredibly heavy, so that the riders needed help to mount and if unseated by the enemy they were unable to run away or defend themselves. Even their horses were equipped with metal plates and trappings to fend off the spears and swords of the enemy.

The horses had their own weapon of attack. In many suits of horse armour the chamfron, a piece of metal across the forehead, was fitted with a sharp spike which was used against an attacker. Naturally the horse had to be exceptionally sturdy even to stand under this great weight, and huge draught beasts were employed. At the best of times slow, these great beasts could barely reach a trot when in battle regalia, so the element of surprise was quite lost. Another important factor was their cost. Such a huge fighting animal consumed vast quantities of expensive grain and fodder and had to be well cared for because of the knight's dependence on it.

All this cumbersome equipment made the horse and rider an easy target for nimble foot soldiers, and the development of cannon and musket finally numbered the days of heavy armour. However, the cavalry was still needed and lighter, faster horses once more became the ideal. In the seventeenth century the King of Sweden, Gustavus Adolphus, revitalised the whole use of cavalry. He trained his men to manoeuvre at speed in small units, and his principles of shock attack and charges with drawn swords led the way for armies all over Europe to modernise their ideas of mounted combat. This organisation was so successful that everywhere cavalry became an important part of the armed forces and remained so until this century.

Counting on the cavalry

In the nineteenth century the cavalry charger was a valuable investment. Armies were becoming professional institutions and the troopers had it drummed into them that their mounts were far more important than they were. The cavalry trooper's duty was to put the health and comfort of his mount on a par with his own – above it whenever necessary. This, however, did not prevent the horrible reality of battle. The cavalry horse was drilled to consider itself safe in formation with its comrades. If the rider, its guiding god, was shot from its back, the charger would almost always try to force its way back into the ranks, quite terrified and often unconscious of terrible wounds. Much the same kind of behaviour is seen today in great races like the Grand National, when riderless horses stay in the race and imperil other riders by the very success of their own training.

Generally speaking, the cavalry fell into two main categories: heavy and light. Heavy cavalry units were intended to act as shock units, breaking the enemy line by the momentum of their charge on strong, heavily-built horses. The light cavalry, on swift horses and with light weapons, would then take up the running and pursue the retreating and disorganised foe. The lancer was the epitome of the heavy cavalryman, heir to the armoured knight of the Middle Ages, whilst the typical light cavalryman was the hussar with his sabre. Both were reminders that the first horsemen usually fought with sword or spear and eventually emerged as a military force in their own right.

During the First World War the cavalry was still much in use for strategical reconnaissance, as protection for the army front, and for work with the infantry. Their dashing charges and manoeuvres covering the advance and withdrawal from Mons, and their work in Flanders, showed the reliability of the British cavalry units and their great value in warfare.

Even with the vast steps forward in the development of mechanised war-machines, horses were still needed in the Second World War. The final action seen by British cavalry troops was part of the Syrian campaign of 1941–42. Mounted troops armed with rifles and swords successfully charged the enemy. It is interesting to note that many of the horses used in this campaign were not the trained and seasoned warhorses of earlier times but English hunters requisitioned by the army. The German army took along a full cavalry division when it invaded Russia in 1941, and the Russians themselves used mass Cossack cavalry forces, armed with sabres as well as sub-machine-guns. Renegade Cossack horsemen served with the German army and, as in 1918, one of the last actions of the Second World War in Europe was a cavalry charge.

Overleaf. *The charging cavalry even finds a place in twentieth century combat.*

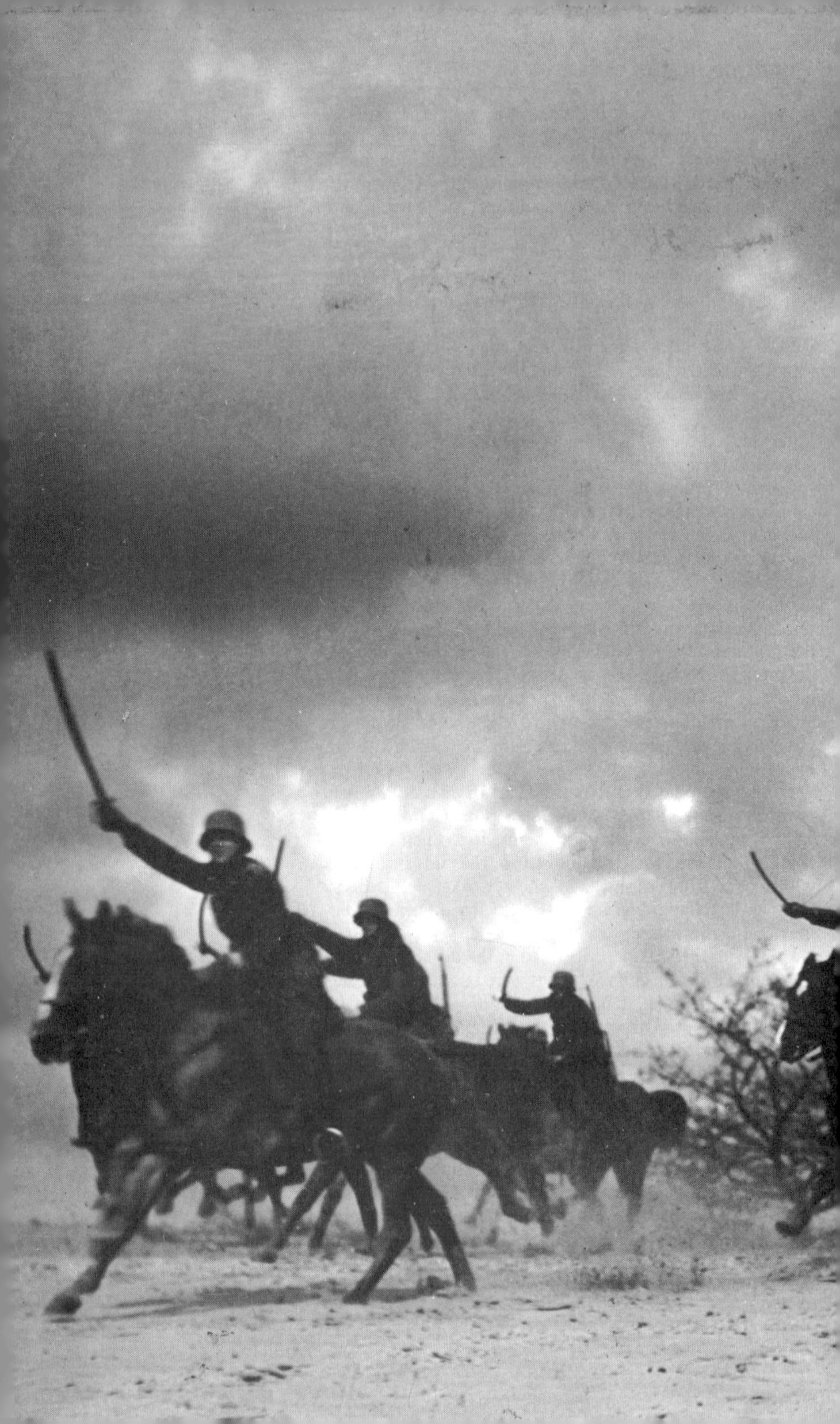

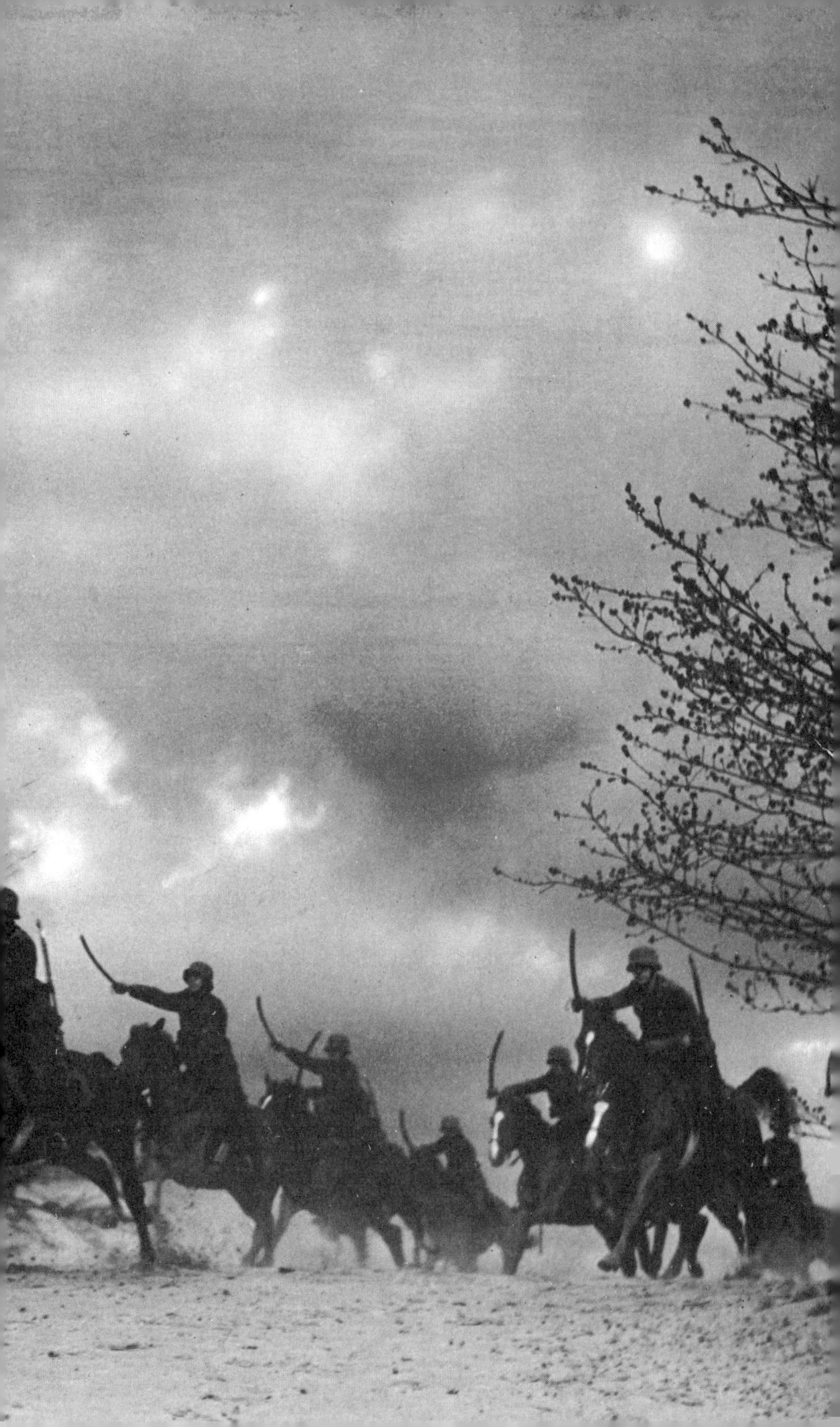

Help from camels

Horses, oxen, mules, camels – these formed the backbone of army transport columns for centuries. The latter two animals had a duel rôle: they were also used as cavalry. Lawrence of Arabia provides some of the best eye-witness accounts of their contribution to the success of the Arab Revolt in 1916–18. He writes: 'Maulud, the fire-eating A.D.C., begged fifty mules off me, put across them fifty of his trained infantrymen, and told them they were cavalry. He was a martinet, and a born mounted officer, and by his spartan exercises the much-beaten mule-riders grew painfully into excellent soldiers, instantly obedient and capable of formal attack.'

One of Lawrence's desert warriors. The camel played a very large part in the Arab Revolt of the First World War.

Opposite. *In this 19th century painting Joan of Arc is depicted astride a huge white charger, leading her army against the crumbling defences of the English.*
Overleaf. *Russian cossacks with their strong swift horses from the Caucasian steppes proved their worth as light cavalry during the Second World War.*

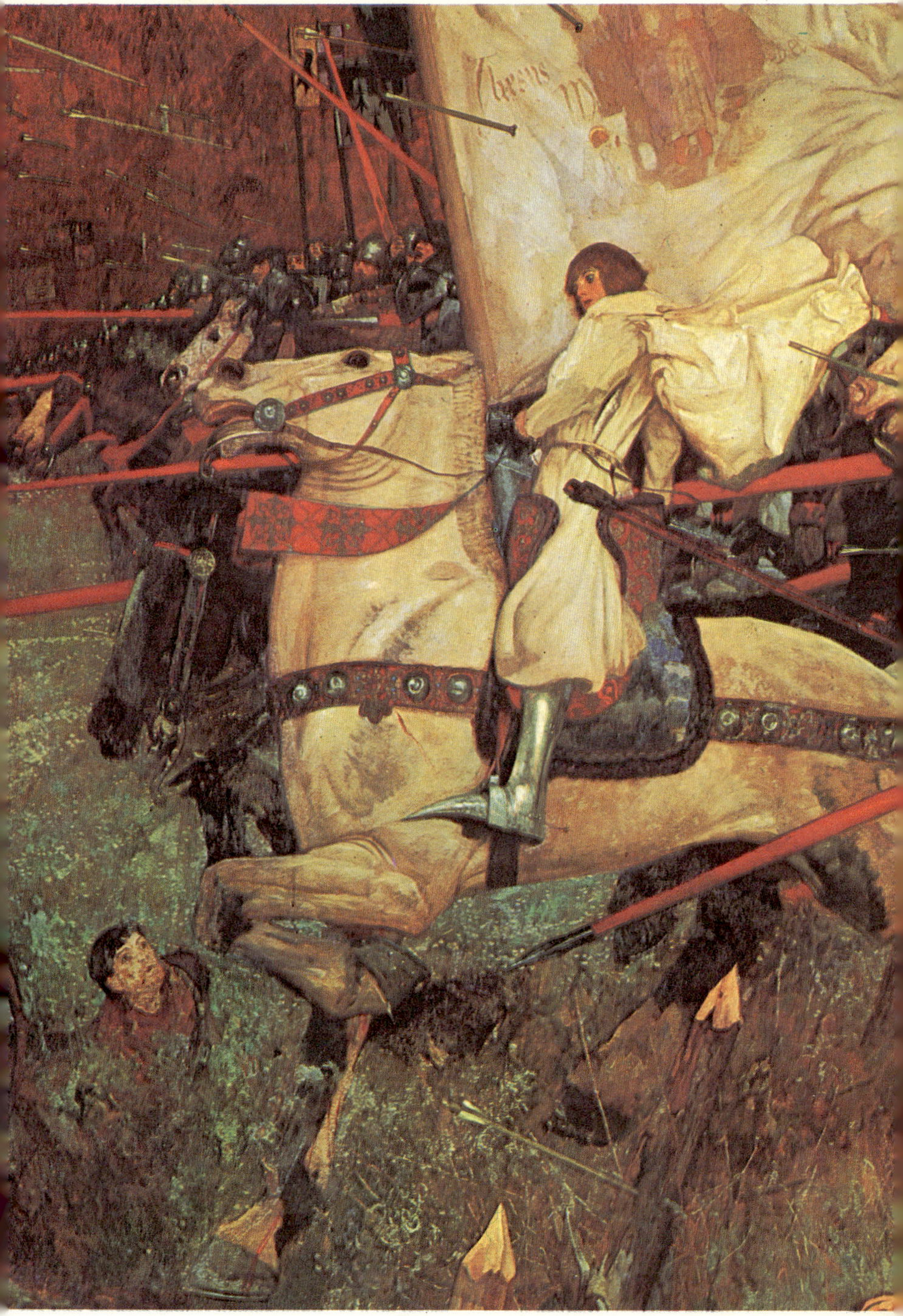

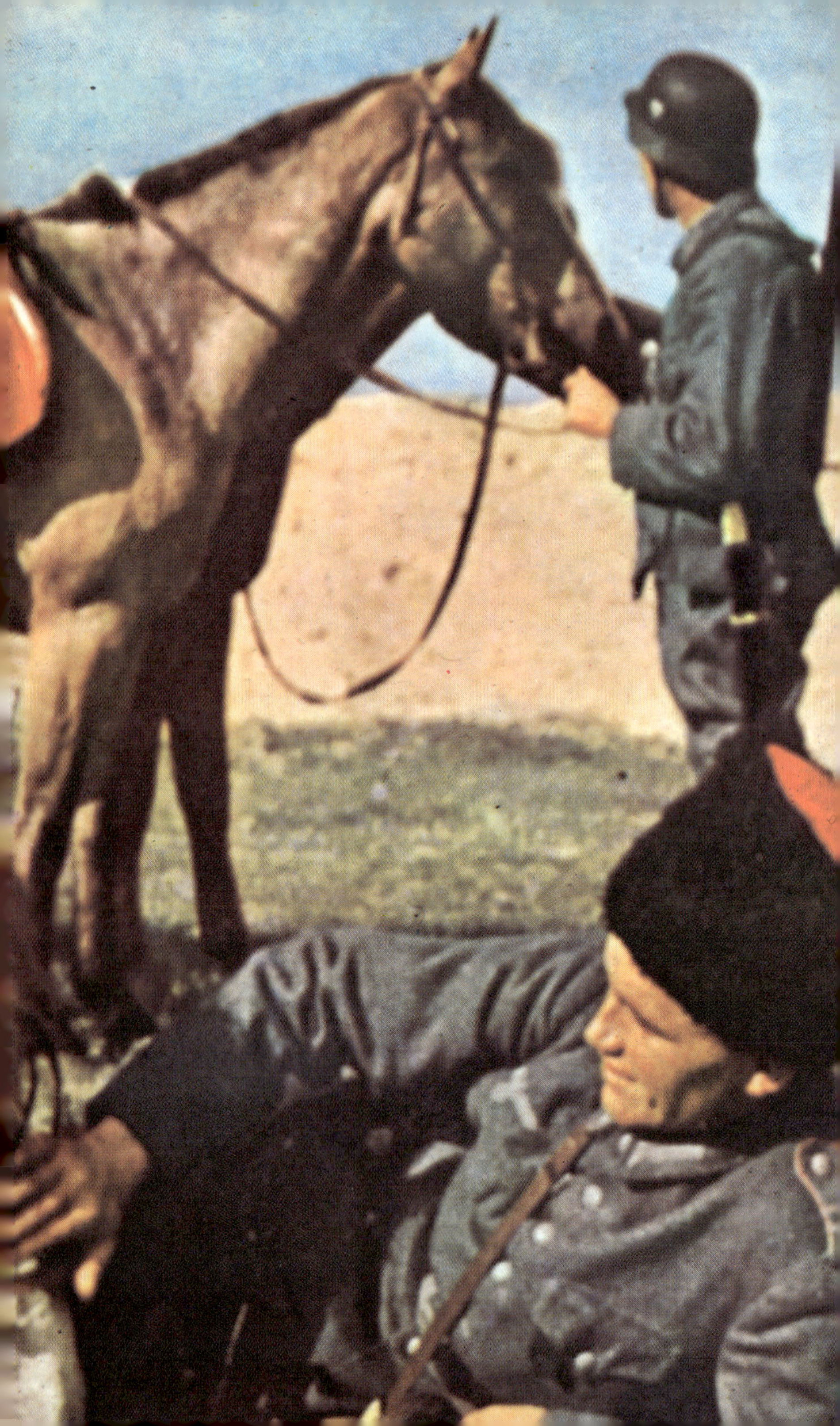

Camels, however, were the mainstay of the Arab Revolt. Thanks to their extraordinary staying-power, Lawrence and his Arab irregulars made incredible camel-borne expeditions in their war against the Turks. Camels were first used in warfare as early as 190 B.C. They proved their worth in the desert battles of the Syrians, who sent their archers off on camels, giving them both mobility and a fine vantage point from high on the camels' backs. Eastern nations have used camels for centuries, and in 1722 the Afghans even went as far as to invent a cannon, light enough to be carried by a camel, which could be fired from the back of a sitting beast.

Apart from their cannon-bearing properties camels are ideal for desert warfare. They are disagreeable creatures when working, but can carry a heavy load for long distances without food or water. Camels can cross territory with deep sand and survive heat and sandstorms that would completely halt a horse or mule train. Despite this hardiness thousands of camels have died during war because they have been forced to stand with heavy loads still strapped to their backs, or have been given insufficient time to graze. They are also useless in the cold. The British tried to use Arabian camels in the Crimean war in 1854, but they could not stand the low temperatures and many died.

When well cared for by its drivers, the camel is an excellent beast of war. Napoleon used camel troops when

In the great heat of the Jordan valley camels proved to be the only successful means of transport for the Desert Mounted Corps.

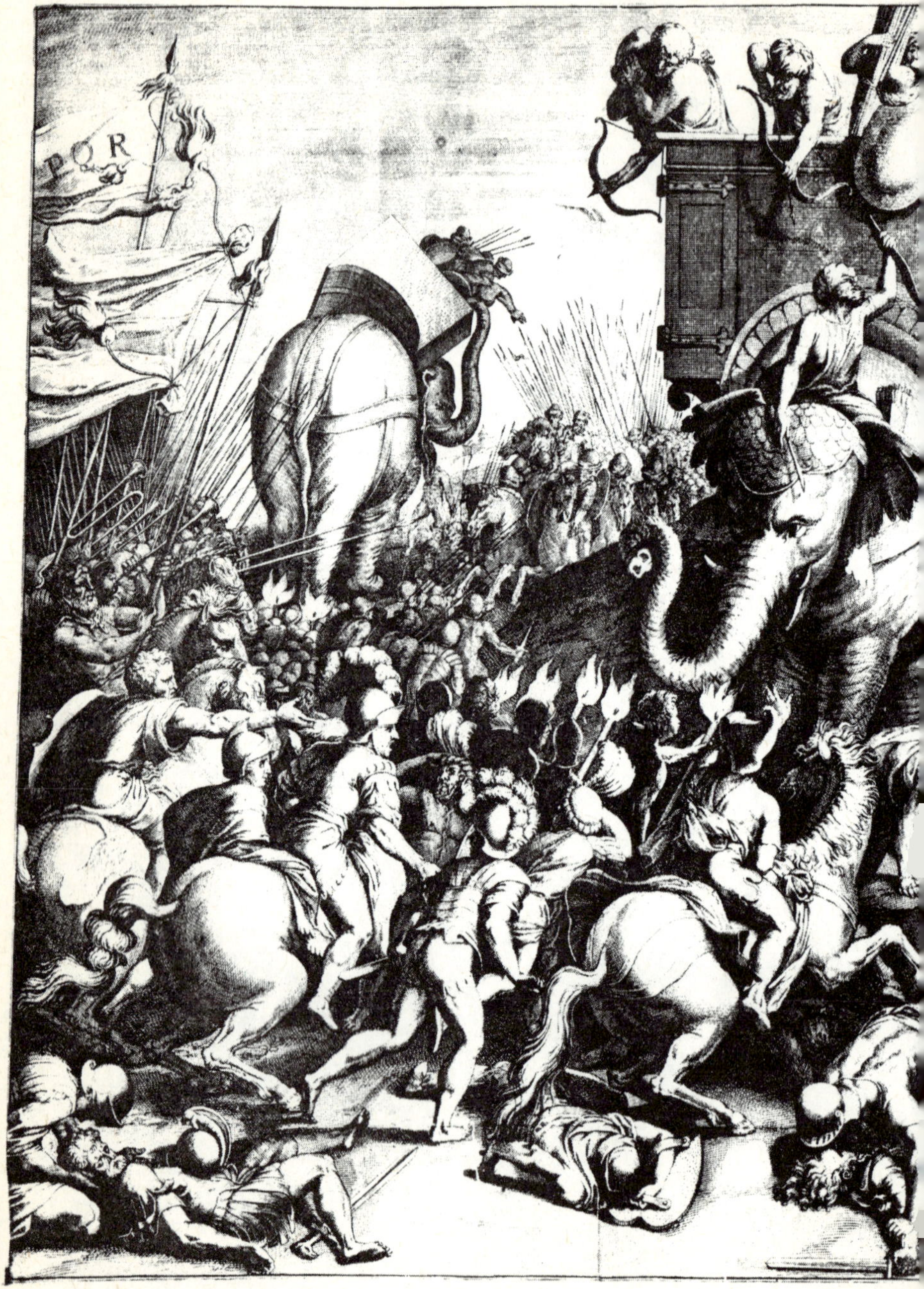
PQR

Despite their size and strength elephants have never really been the successful war animals that might be expected. In an artist's impression however they are pictured as vast beasts, trumpeting their way into battle and causing panic among both horses and men.

he invaded Egypt in 1798 and both British and Egyptian armies used camels in the Sudan campaigns. A dozen or so years later, the Camel Corps were used in the 1914–18 war against the Turks and Senussi.

Large and small recruits

Elephants, too, have long been used both on the battlefield and in the baggage-train, but their performance in battle has never reached expectations. Despite their huge size and terrifying appearance they never seem to have swept a battlefield like living tanks, as they might be expected to do. The great Carthaginian general, Hannibal, placed a great deal of faith in war elephants, faith which cannot be said to have been rewarded by positive results. Hannibal included thirty-seven elephants in the hotch-potch Carthaginian army which he led against Rome in 218 B.C., but only one, which he was supposed to have used as his personal transport, survived the arduous crossings of the Pyrenees and the Alps. The 'Waterloo' of the Rome/Carthage struggle, the Battle of Zama in 202 B.C., opened with a mass Carthaginian 'tank attack' by eighty war elephants. In a battle decided by the horse cavalry of the Romans and their Tunisian allies, the elephants' only success was in roughing-up the lightly-armed skirmishers

Porthole for pigeons

A carrier-pigeon is released from a special hole in a tank. Pigeons were used frequently as messengers often carrying vital information from the Front. This one was photographed near the Somme towards the end of the First World War.

122

screening the Roman line, before they stampeded off the battlefield.

Elephants cannot be drilled for battle as horses can. They panic, or as Kipling's elephant Two Tails says in *Her Majesty's Servants*: 'I can see inside my head what will happen when a shell bursts; and you bullocks can't.' A reasonable explanation, indeed, for the elephant's fear of battle.

Yet elephants can perform wonders of hard work for the supply-columns, even in modern times. Throughout the vicious Burma jungle campaigns of 1942–45 they were used as bulldozers and heavy-duty transports by British and Japanese alike.

Size and strength, however, are not the only animal attributes that have proved of value to man on the war front. Among the animal lightweights, pride of place is held by the carrier-pigeon, long used as a swift and unobtrusive battlefield messenger. One pigeon which fell dead after carrying a vital message through the hell of Verdun in 1916 was awarded a posthumous *Legion d'Honneur*. Stuffed and preserved for posterity and with full military honours it was solemnly exhibited in a Paris museum. Pigeons have been present on several historic occasions. Superstition did not worry the crew of the first R.A.F. plane to torpedo a German pocket-battleship at sea. On June 13, 1941 – Friday the thirteenth – the crew of 'W for Wreck' No. 42 squadron were carrying Carrier-Pigeon Pannier Number Thirteen. They were in fact the only crew out of the thirteen that left their base which managed to put a torpedo into their target for the night, the pocket-battleship *Lützow*.

In November of the same year a South African tank commander in the Eighth Army had a messenger-pigeon alight on his shoulder on the eve of a battle with Rommel's Afrika Korps. Wondering if the bird was carrying the fate of the Desert War on its leg, the commander unfolded the message with bated breath – only to read an extremely rude version (in English) of 'Damn you, leave me alone'.

The horror of the trenches

Some of the most unfortunate animals ever pressed into military service were used in the trenches in the First World War. Mine-tunnels were thrust forward by both sides in order to plant heavy charges of high explosive under sections of the enemy trenches and blow them up. Gas was used to discourage such mining activity, and caged canaries and mice soon formed part of the tunnelling companies' regular issue. The cages containing the mice or canaries were held out in front of the tunnellers, and because the animals were quickly overcome when gas was present, they provided a ruthless but effective anti-gas early-warning system.

The dog, too, entered the trenches. Perhaps his most valuable rôle was the gruesome one of rat-killing. The carnage on the worst sections of the Western Front produced a rapid development of a terrifying form of rat which acquired the taste for human flesh from the abundance of corpses in No Man's Land, and which frequently made life unbearable for the men in the trenches and dugouts. Tough terriers were invaluable allies in the war against the rats, and often ran up individual scores larger than the tallies of enemy planes shot down kept by fighter aces.

Dogs had other uses too. Some armies used them to tow machine-guns; elsewhere they were trained to lay field telephone-cables across No Man's Land, carrying miniature reels on their backs. This in particular was an ideal rôle: a dog is a much smaller target than a man, and anyway is far easier to sacrifice than a fully-trained soldier. But perhaps the most dramatic use to which dogs have ever been put in modern warfare was as anti-tank weapons by the Russians in the Second World War. In the desperate months of 1941–42, before the Nazi invasion had been held at Leningrad, Moscow and Stalingrad, there seemed no way to stop the Panzer tank divisions. Impromptu flame-bombs such as the famous 'Molotov cocktail' were used, and also the rather hideous concept of

At the beginning of the First World War dogs were used by the Belgian infantry to drag machine-guns.

One-time operation

Russian soldiers with 'mine dogs'. These dogs were trained to dash under German tanks and stay there. The timed explosive charges strapped on the dog's back went off exterminating the tank and the dog, the idea being that it was better to sacrifice a dog than a trained man.

the mine dog. These canine mines were trained to run under tanks which had their engines running. They were then taken into the front line and released towards the German lines with explosive charges strapped to their backs. These had contact detonators, and the idea was for the dog to immolate itself under a tank and – the Russians hoped – to take the tank with it in the explosion.

Well before mechanised warfare, too, the dog had played its part on the battlefield. In the nineteenth century one South American revolutionary went into battle with a disciplined but formidable pack of Newfoundland dogs, trained to savage dismounted enemy horsemen.

Today, the army still keeps horses but they are mainly for ceremonial purposes, but tracker dogs and dogs trained to take weapons from a potential enemy are still used in many armies of the world. In civilian strife dogs and horses play an increasingly important rôle each year. Dogs are now being trained to track, trap and catch criminals, and their strong sense of smell is used to detect items such as dangerous drugs as well as the more traditional searching for corpses. The police horse plays a valuable part in keeping the peace in large crowds. The horses, with their big feet, command respect from even the most unruly gathering and they are used as living weapons to push back crowds and demonstrators when things begin to get out of control.

Further ways in which man's ingenuity may harness the animal world are uncertain, but there are many possibilities which will no doubt receive the attention of warlike man in time. For example, whales, porpoises and even large shoals of fish can make contacts in the way enemy submarines do with modern detection apparatus. Flights of geese can appear on radar screens like hostile aircraft. Even medium-sized birds colliding with jet aircraft can do enough damage to knock the aircraft out of the sky. It may well be only a matter of time before these sources of living weapons are tapped by men in the desire to increase their power over other men.

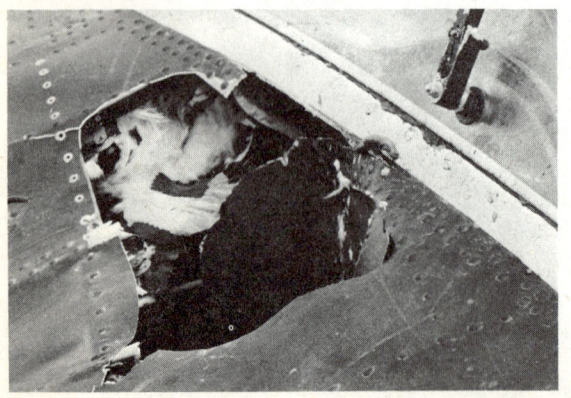

The large hole in this Beverley transporter was caused by a bird, probably a seagull, which can be seen inside the hole. The Royal Air Force is most concerned by these collisions which cost £1 million every year in repairs to military aircraft. Most strikes are minor but five planes have been totally wrecked and there is the danger of fatal accidents. Prime culprits are seagulls, pigeons, plovers and starlings.

Man – what went wrong?

History saw many cruel leaders, vast armies, slaughtered peoples, much suffering & grim death. You must each free yourself from leaders, passports, territories, & all barriers which seek to divide,

from A letter to a few good poets I know by Dave Cunliffe

Murder, suicide, crimes of violence, civil wars, world wars – man is not only unique for his intelligence but also for his cruel nature. In warfare all respect for human life vanishes and man kills his own kind without inhibition. Man's aggression is out of hand and cannot be likened to animal war.

The devices which the human mind has conceived for the torturing and killing of fellow men form a catalogue of horrifying brutality. Man's status in the animal world is unique. He is the most cruel animal, and the only one deliberately to kill his own kind.

Cruelty in warfare is not only a feature of past history, to be put down to the uncivilised attitudes of our ancestors. It continues to grow in magnitude, as the Nazi programme to exterminate twelve million Jews in the Second World War shows us. At a conservative estimate, murderous quarrels and wars were responsible for fifty-nine million deaths between 1820 and 1945. Throughout his written history man seems consistently to reject the advantages he has over the animals. How can any creature capable of foresight, detailed memory and correlation of thought – any creature of intelligence – keep up an insane record like this? Much of the answer could lie in the origins of man, and the relationship of his behaviour to that of the animals.

War and lack of respect for human life seem to be ingrained in man's society. Few parts of the globe are without warfare of some kind or another, and to kill many foes in battle continues to be considered one of the most glorious achievements a man can aspire to. Violence is a favourite subject for entertainment, too, if the popularity of westerns, thrillers and war films is anything to go by.

Student riots: this time in Greece, but the scene might be anywhere from the United States to Japan. Violence is playing an ever increasing part in human life.

Even children's games delight in mock killing and wounding. From the moment we are born we are subjected to a stream of violent events going on in the world around us, if not within our own experience then brought to us on the television screen.

Clearly, civilisation may be obscuring the true situation with respect to the war-like nature of man, so there has been much study of less 'advanced' society. Primitive tribes the world over contain many bizarre expressions of cruelty and aggression. Mutilation of the body is common for religious reasons and for initiation ceremonies; head hunting and cannibalism still exist; sacrifice has only recently become transferred from people to animals. In fantasy too, man is violent. His various mythologies centre round aggression, and the hells described by religion are places of frightful cruelty and torment.

A recent study by an expedition from Harvard University of a particularly war-like tribe in New Guinea describes the lives of the people as 'an unending round of death and

This is the kind of mass destruction which it is now in man's power to let loose. On 6 August 1945 Hiroshima gained the distinction of becoming the first city to be destroyed by the atomic bomb. Since then the nuclear arms race has grown apace.

revenge'. In them, it has been suggested, we observe man in his original, wild state. On the other hand, there are peoples to whom war is completely foreign, such as the Eskimos. Nevertheless, some students of human behaviour have argued that the fact of all this violence points to an inborn tendency to kill, an over-aggressive 'drive' inherited by every human being.

Is culture the culprit?

This conclusion is far from being inevitable, however. In fact, in the light of our knowledge of the process of evolution it fails to make sense. All evolutionary changes take place because the changes are of advantage in promoting the survival of the species. A drive towards killing members of the same species cannot have become built in to the human make-up by any normal process of evolution, because it is so obviously a disadvantage to the species. If all time and energy is devoted to killing each other, how can reproduction and the maintenance of numbers occur? Behaviour which is the product of biological evolution is adaptive however strange it may appear to us. For instance, we have seen that when a female mantis eats her mate there is a purpose in her seemingly unpleasant activity. She is not endangering the survival of the species as her mate has already served his purpose. War, in contrast, is non-adaptive.

So the opponents of the 'built-in' idea suggest that war is not an inherited tendency in man but a product of his culture, learned and not inborn. Culture is the knowledge and attitudes of mind which growing children derive from parents and other people they meet or associate with as they develop. Men, women and children learn from schooling, reading, general upbringing and the political climate around them. Culture includes religion, the language spoken, the way of dressing and eating, the taste in art, and ideas of freedom. Because culture surrounds the human baby from birth, it is almost impossible to disentangle its effects from those of heredity.

One unfortunate aspect of human culture is that it separates people into groups with very little in common. Communist and capitalist states, for example, share few ideas, and when language is also a barrier to communication misunderstanding grows apace. Even within a country there are groups which do not understand, and so mistrust, each other. Social classes separated by income, education and even accent add further to the division. The old do not understand the young and vice versa. The world is divided into 'them' and 'us', and if people do not behave as we expect then we are afraid of them, and separated from them.

Human parallels

Poison fangs, lethal claws and sharp teeth are just three kinds of animal weapons. When an animal comes into contact with an enemy then it will use its weapons to kill. However, in a rival encounter the same animal will use the same weapons but only as a means of threat. Polar bears cuff each other with their hefty paws and bite at each other's fur but they do not use these deadly weapons to the extreme. Rattlesnakes die from their own poison but death is not the purpose of rival battles and so they fight in a stylised manner according to strict rules. The only creature to use weapons to kill his own kind is man. With few inhibitions a soldier will train a gun on a human target and fire.

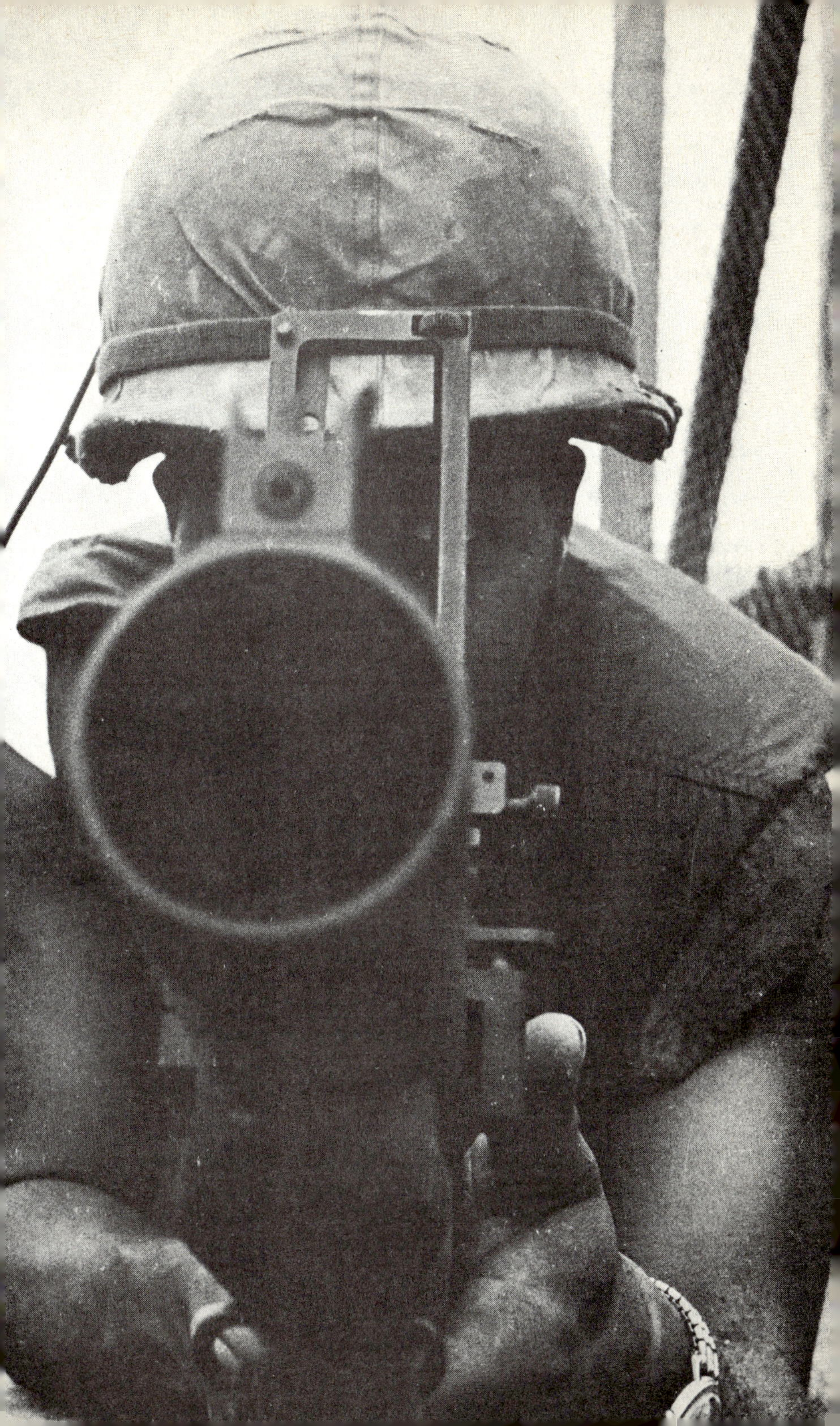

War may have arisen because groups no longer understood each other. Rituals of communication which keep aggression under control, including the important gesture of submission, are of no use unless the whole species understands them. All baboons 'speak the same language' because they all recognise the meaning of another baboon's behaviour. The same cannot be said of mankind. In a sense, human groups have come to regard each other as different species, so that all the rules of restraint are lost and attack takes place without inhibition just as a predator kills its prey.

Problems of early man

When did all this begin? To find out how man came to differ from other animals one must piece together fossil evidence for human evolution, painstakingly collected from many sources. The record is far from complete, and there is no unanimity on how it should be interpreted, but a good place to begin is with the 'man-ape' *Australopithecus* which lived in Africa over a million years ago and is probably the forerunner of man.

The ancestors of this creature were probably tree-dwelling primates with a varied diet and flexible behaviour, rather like present day chimpanzees. Then about twenty million years ago they moved from the forests to more open country, and by the time *Australopithecus* arrived on the scene several important changes in behaviour had taken place. A diet of mainly fruit was replaced by one in which meat represented a large proportion. Even more significantly, *Australopithecus* was able to walk upright, so his forearms and hands were freed for other tasks, notably the making and using of stone tools and weapons for use in hunting.

The danger in *Australopithecus*' tool-using lay in the relative speed with which he acquired these weapons. There are many predators which carry as parts of their anatomy far more formidable weapons than sharpened stones. But they rarely use them against each other and they do not kill deliberately. The mechanisms of threat, ritual fighting and appeasement which prevent deaths among these animals have been described. The rituals evolved alongside the weapons. Man's transition from a peaceful existence in the trees, however, was relatively sudden. He had not taken the millions of years that lie

Opposite. *May Day parades in Moscow are not solely for Russian consumption – they are intended to impress and warn the rest of the world of the USSR's military might.*
Overleaf. *The hydrogen bomb is considered to be the ultimate threat. Tests on Bikini Atoll in the Pacific show the holocaust that results from use of this weapon.*

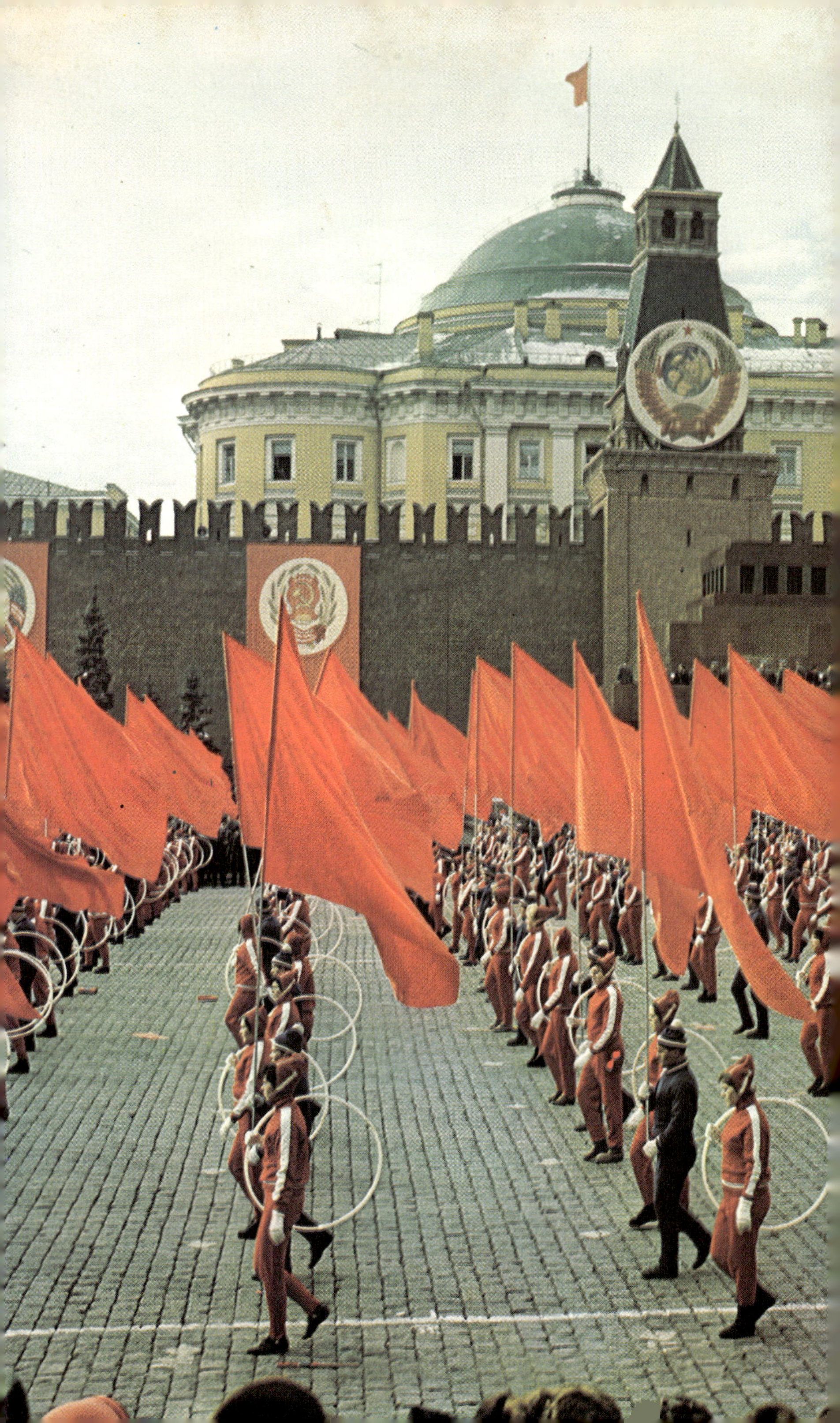

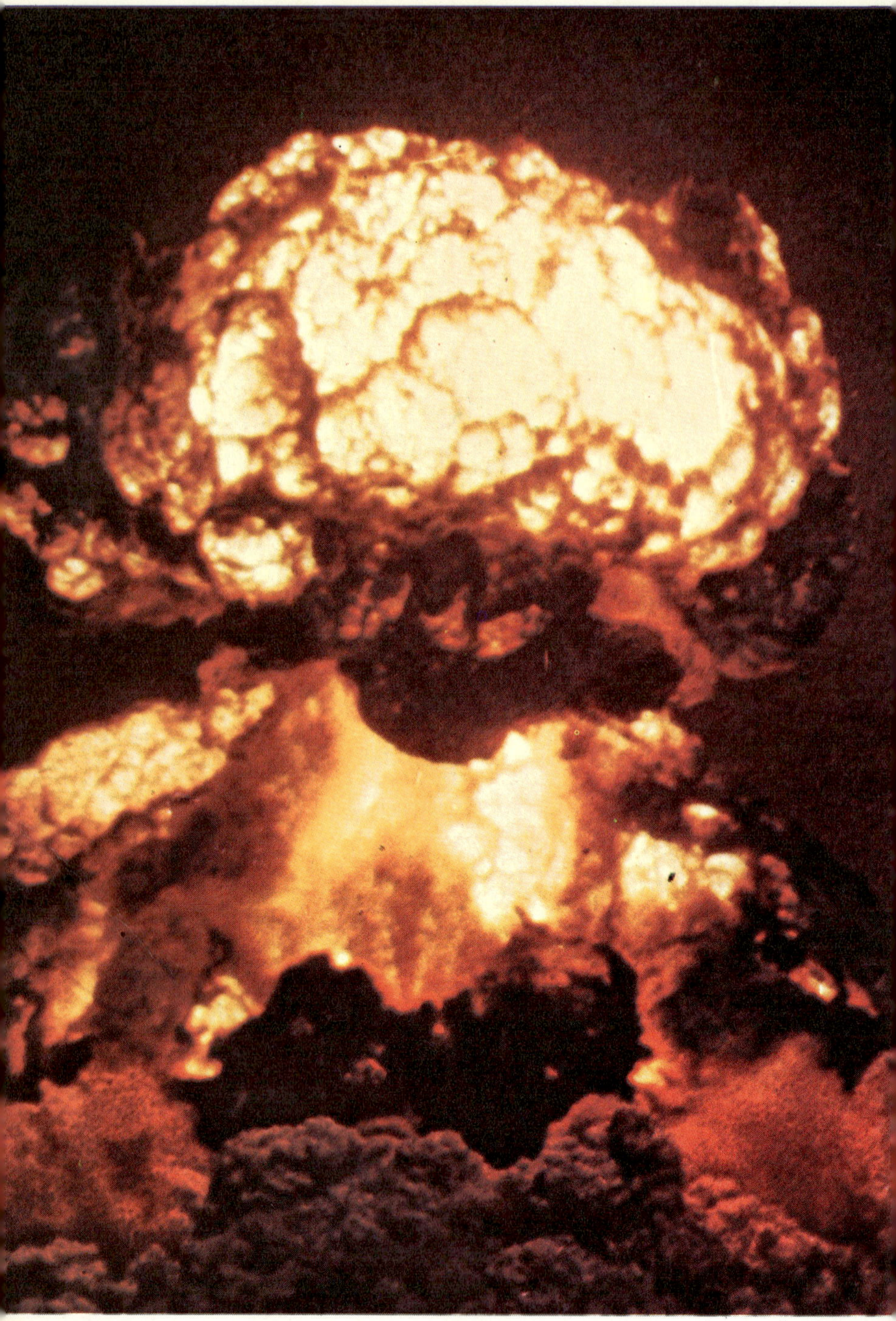

behind such weapons as wolves' jaws and snakes' fangs to develop his stone tools; he had not evolved the very necessary inhibitions that should have accompanied them. He was, and possibly still is, an uninhibited ape armed not only with elementary tools, but with newly evolved intelligence and memory with which to develop them.

The change from fruit-eating to hunting had a profound effect on behaviour. No longer did family groups simply forage for food, with little or no territory. Groups of men now co-operated to hunt game and some division of labour was inevitable. Pregnant females and those with small infants, old males and youngsters made poor hunters,

No-one can ever know what our man-ape ancestors really looked like, but this famous imaginative reconstruction of Australopithecus *may give us some idea of the savannah-living, tool-using hunter.*

The country retreat of a well known millionaire. In his desire for peace and quiet he has ringed the estate with guards, dogs, and fences labelled 'keep out'. Human beings, like animals, need their privacy, and so they keep others at a distance by means of territorial signs.

and larger game required the attention of more than one family group. Thus, over tens of thousands of years, was born the tribal unit. The fittest males hunted, the others stayed at one sheltered spot. This spot was their territory – what we would call home.

So long as the population remained small, there was little competition for food, but after the development of agriculture and the beginning of towns and large settlements there was a sharp increase in the number of people trying to live off the available food supplies. Firm evidence of inflicted death and warfare is first found in the temple towns of seven thousand years ago. Economic competition and division into groups were beginning to have their effect, and the real danger of the lack of inhibition in the use of weapons came into play. In contrast with other animals, the territorial and aggressive tendencies of man are not under control, and can lead to fighting where competition is fierce. On the other hand, simple hunting and gathering tribes such as Bushmen are even today able to live peacefully with their neighbours because they are spaced out over a large area and do not compete for food.

A home to defend

Sometimes man's territorial behaviour is very reminiscent of an animal's behaviour. Animals owning large territories obviously cannot defend all points on their borders at once. Many of them, therefore, leave markers at strategic points. Bears, dogs, deer and many others mark their territorial limits with scent in the form of urine or dung, or with special glands. Modern man's parallel to this is the notice that 'trespassers will be prosecuted' – a familiar 'visual bluff', usually used as a second line of defence. The first line of defence comes from man's ability to build extensive barriers. Many country estates, far too great in area for efficient patrolling, are bounded by fences, thick thorn hedges, and walls topped with broken glass.

On a smaller scale, houses and cars are territories, each marked with some distinguishing feature to separate it from the rest. By every means from brightly-painted front doors to elaborately trimmed hedges, the houses and gardens are set apart from their neighbours. Many people decorate their cars with mascots, stickers, brassy exhausts or useless cushions and ornaments; the owners carry their own personal territory, suitably labelled, with them. The idea of car-territory can be carried farther, so that amusing experiments are possible. It has been suggested that a motorist stopped by the police should get out of the car and go to meet them. At the same time he should adopt a rather 'hangdog' expression to indicate submission. Thus the offender, instead of remaining in his territory – which would be an aggressive gesture – is humbling himself in the hope of placating the police, as well as making suitable 'you are dominant' gestures. This may sound ridiculous, but nowhere is man more territorial than in a car. Minor races, tight-lipped battles of nerves by the most inoffensive-looking people, and straightforward aggressive bellows on the horn are everyday events on any crowded road. The results can be read in accident figures which make some of the war casualty lists of history look insignificant.

Territory is a major element in many sports. Spaces are allotted to each side and the object is invasion and defence. In games such as football and hockey, individual players are given the areas of ground best suited to their ability; to attack or defend, to play from left or right. With rules to prevent actual violence – in theory at least – the players indulge in a ritualised war for the benefit of spectators who derive vicarious pleasure from identifying with each side. It is significant that, like animals on their own territory, teams have consistently better results on home ground.

Sport may well be a valuable outlet for aggression which could otherwise be expended in more serious fighting. It also has value in bringing peoples of widely differing cultures together with a common interest, being a ritual that all can understand.

Rituals of many kinds play a part in human life, but they are culturally, not biologically, determined and so not strictly comparable with those of animals. Men and women also have appeasement gestures; for instance shaking hands is a means of overcoming the initial distrust, or shyness, between individuals. Bowing or kneeling as a gesture of submission to king, queen or god is still world-wide. In our courtship, we revert to childish gestures and language in exactly the same manner as, for instance, gulls. This ritual overcomes any aggression – or fear – which may lie between a pair in close contact.

It is, however, when men actually fight that one sees the closest parallels with animals. Inside or outside a bar in any of the world's cities, battles for dominance or disputes over women are common. Like tom cats howling or stags roaring, the men shout at one another, and, animal-fashion, the fight can stop at this bluffing stage. The contestants stand in threatening postures, fists clenched, and, if one is particularly confident, he will stand almost nose to nose with his prospective opponent, deliberately disturbing him with unnatural proximity.

The fight itself, once started, can consist of anything from futile pushing – like stags and sheep – to blows from the fists, head and feet. If the fight is between only two men, weapons seldom appear. Whether their absence is due to fear of repercussions from the law or to the remnants of inhibition is a matter for conjecture. Normally men carry weapons as a threat rather than for actual use. It would seem that, consistent with the animal procedure of the fight, we retain some inhibitions about personal murder.

Of rats and men

Yet murder, suicide and crimes of violence increase almost daily in the more developed countries. People seem increasingly aggressive in everyday life. A look at road accident figures or observation of the pushing, elbowing irritability of a bus queue confirms this. The reason is almost certainly overcrowding.

Murder and suicide certainly represent abnormal behaviour, but perhaps the number of abnormal people is increasing because of the stress and crowding of modern civilisation. Only when they are overcrowded or in captivity will animals lose their normal behaviour patterns, even to the extent of killing one another consistently and deliberately. An account of an experiment with rats presents some grimly familiar results. Four pens were arranged end to end. The two end ones had single entrances to the two middle ones, and the latter were connected by many entrances. Fifty rats were put in, liberal food and water supplied, and the animals allowed to settle down in whatever way they chose. Predictably, the two most dominant rats took up residence in the end chambers, keeping out competitive males and establishing family units. When the young became mature, they were forced into the middle pens.

The overcrowded middle pens were rodent chaos. All

In an overcrowded enclosure, the territorial fights of male rats become unnaturally aggressive. It is tempting to draw parallels with life in human cities, where stress may be the underlying cause of increasing violence.

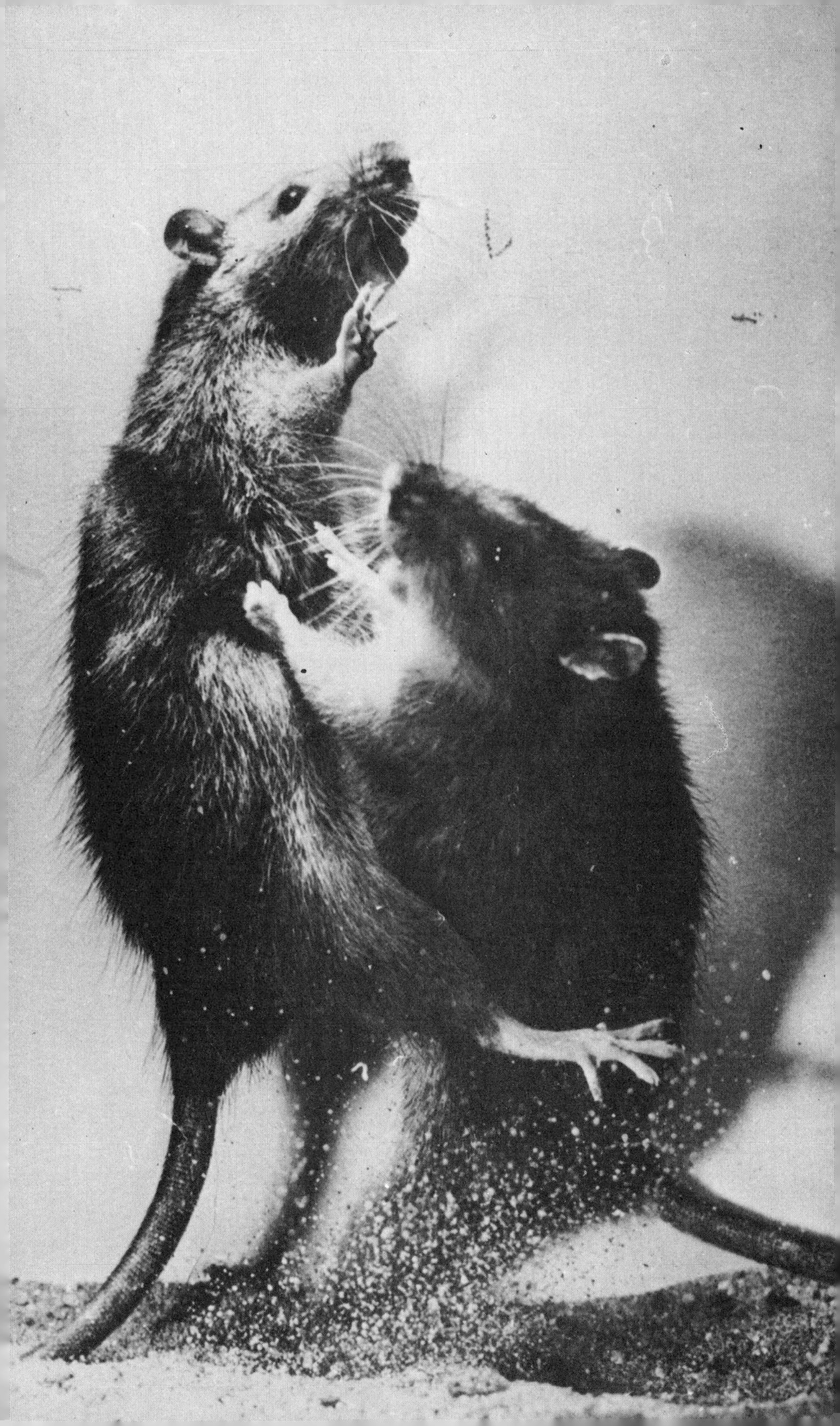

order and hierarchy was lost, and even the sexual patterns were abandoned. All semblance of courtship ritual disappeared; rape and homosexuality were commonplace, and gangs of perfectly well-fed rats preyed on weaker members of the community, or stole food from them. Infant mortality rose to ninety-six per cent as mothers ceased nest-building and made no attempt to rear the young. In fact, many families were eaten straight after birth.

Similar changes seem to be happening in our societies. Baby-beating and infanticide are at their highest in overcrowded slum areas. The gangs of marauding rats may find their parallel in the skinheads, hell's angels and thugs of the towns, at least insofar as they are a product of society's ills. At a slightly less violent level, everyone forced to live in crowded conditions is under a constant strain, leading to abnormal behaviour. When the strain becomes too great, an ever-increasing number of people take their own lives. The current theory on suicide is that aggression turns inwards, causing the kind of hate one would accord one's bitterest enemy to be directed to the self. The enemy within is carried around day and night, and destruction is the only way out.

The strain on people has a fairly simple derivation: proximity. As shown in chapter two, territorial animals in flocks keep a consistent minimum distance away from one

Left. *One of the greatest problems created by the population explosion is the lack of adequate housing for all the millions of city dwellers. A view of Hong Kong shows, in the foreground, the shanty homes typical of so many towns with a high proportion of very poor people. Behind are crowded apartment blocks where hundreds live in close and uncomfortable proximity. New blocks rise in the distance to take the overspill, and so the sprawling growth of the city continues, hand in hand with increased stresses on the inhabitants.*
Right. *The early 1960's saw a new manifestation of the discontent of British youth – the battles of 'mods' and 'rockers' on seaside beaches, which caused a widening in the gap of understanding between the generations.*

another. As the proportion of people living in towns climbs steadily, so people are forced into the close, almost intimate company of strangers. Were they animals in the wild, they would soon migrate, or like lemmings or sika deer, suffer some internal upset and bring about a crash in population.

Man has reached a critical point in his overcrowding. He must either limit the population and distribute it more evenly, or wait for the social and economic crash that will be the result of allowing it to continue its present expansion. If we do retain some lemming-like mechanism released by crowding stress, then it will not be long before we see its effects – if the current situation is not symptomatic of it already.

The 'uninhibited ape' at war

Captivity can provoke abnormal behaviour in many animals. Some become neurotic and irritable, others lose any sexual discrimination and mate with anything of approximately the right size and shape. The majority, particularly the predators, show behaviour that can only be described in human terms: boredom. They have nothing to hunt, nowhere to release their wandering and hunting

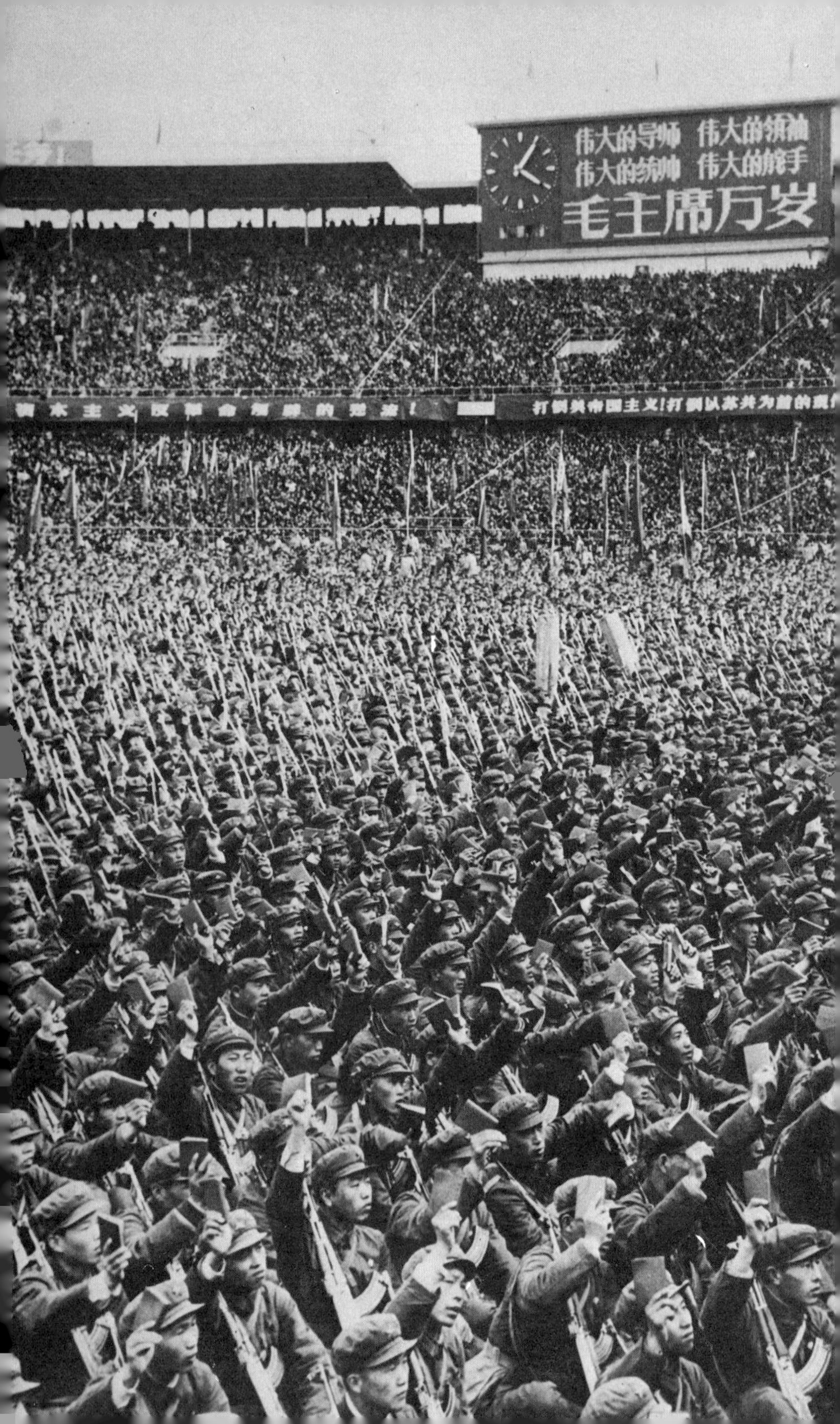
伟大的导师 伟大的领袖
伟大的统帅 伟大的舵手
毛主席万岁
打倒美帝国主义！

instincts. Man could be in much the same position. He has tamed nature; instead of having to adjust to his surroundings, he adjusts them to suit his ways. He has made himself captive within his own world, and has left few outlets for aggressive or hunting tendencies.

War has more effect on the boredom of life than on the population numbers. With the exception of France in the First and Russia in the Second World War, population has always recovered quickly from the apparent carnage. On the other hand, it has bound up the energies of whole nations, turned their attention outwards towards the opposition and upwards to the leaders. Here is an animal situation on a massive scale. The hierarchy of rank is never stronger than in war. In the armed forces it is especially obvious, even to the ritualised appeasement gesture that is a salute. To all appearances, the people of a threatened country are like a group of animals reacting to a potential intruder on territory.

The parallels do not cease with this. Whole nations, represented by their leaders and the armed forces, go through rituals of display and threat. The speeches of politicians, the delivery of notes and ultimata, even the manoeuvring of armies all fall into the category of ritual threat. World war has, in fact, reached the pitch of sheer bluff. The weapons we now possess are, hopefully, too destructive for any intelligent person to use. So, like a stickleback turning its red sides threateningly towards an intruder, like a lizard erecting its frilled collar, Russia parades her military might at the May Day parades, and both Russia and America shoot rockets into space to show off their technological and aggressive capacity. The space race must surely be the most expensive threat display there will ever be.

Left. Waving their copies of Chairman Mao's 'thoughts', Chinese workers demonstrate their solidarity at the inauguration of the Peking Municipal Revolutionary Committee. Large groups of people conditioned into believing their leaders implicitly, and convinced of the evil of other forms of culture, make formidable military forces.

Below. At military parades, traditionally held on May Day, Russia parades her missiles and armies in an effort at international intimidation.

Overleaf. Another mass gathering, this time a Nazi rally addressed by Hitler in 1938. A powerful speaker, he rapidly achieved the status of a demagogue and, playing upon the group instincts and aggression of his listeners, succeeded in leading the German people into war.

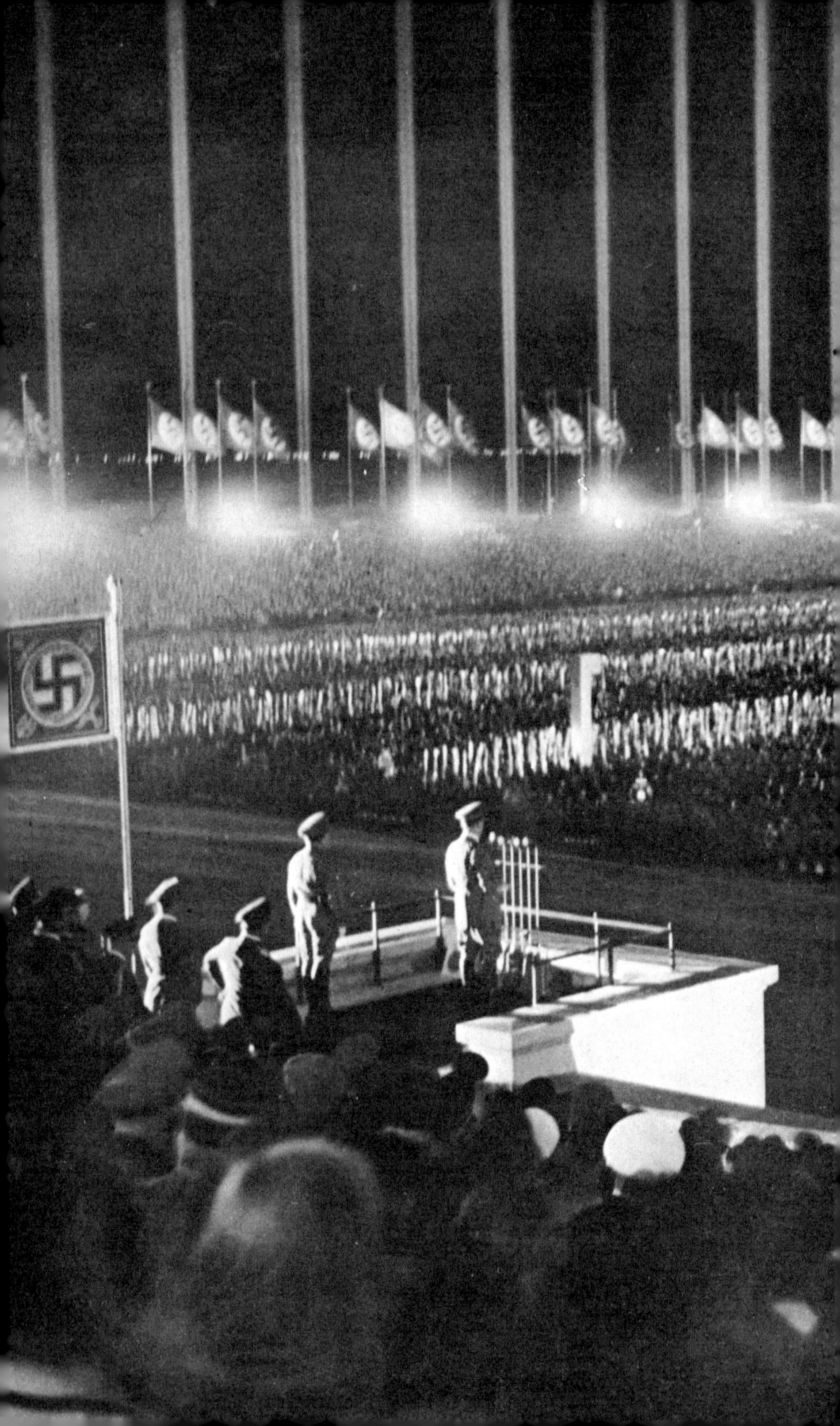

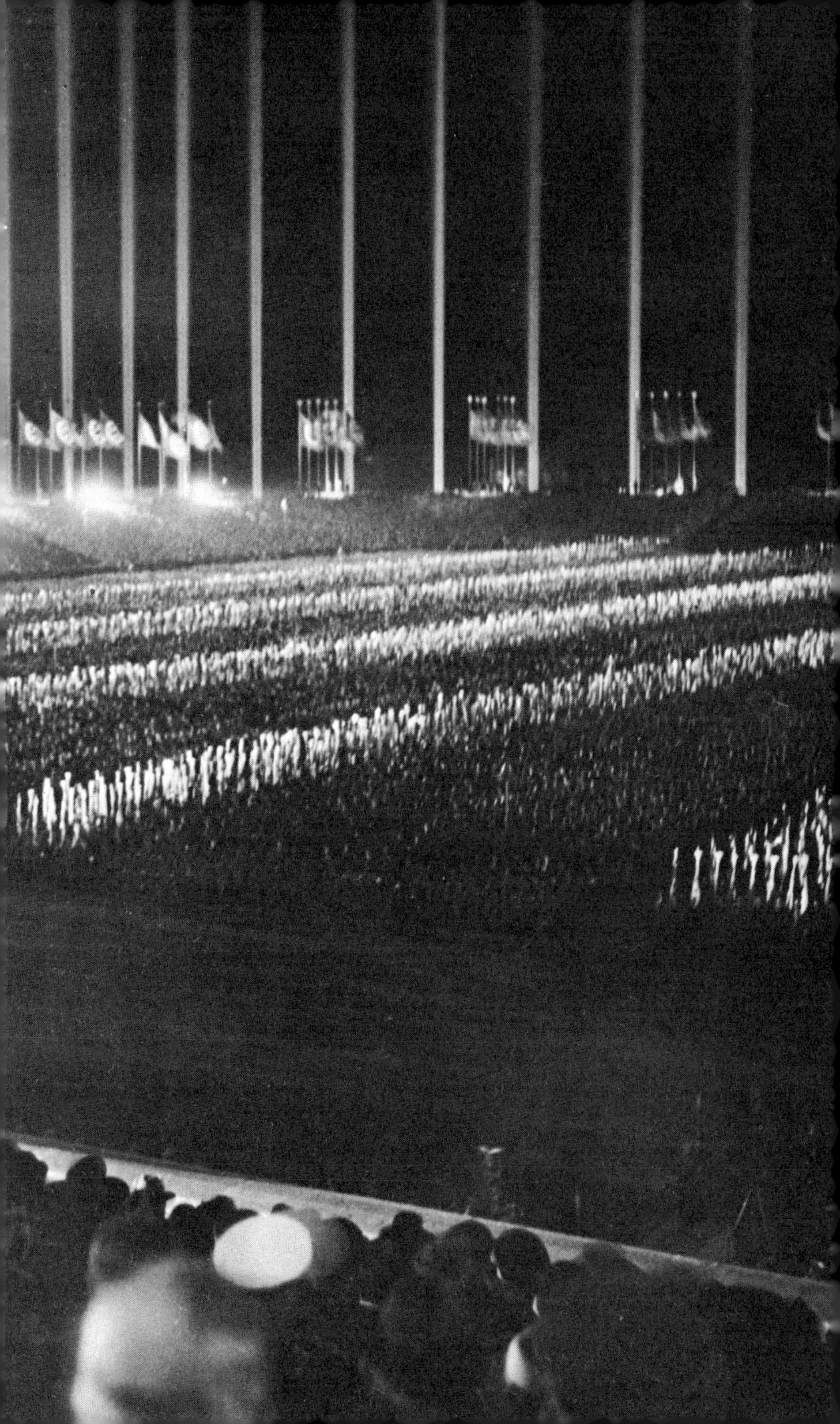

Outlook for the future

In modern warfare life is very cheap. Any inhibitions a man may have about killing at close quarters – as in the street fight – are completely overcome by distance. To the artilleryman, human lives are no more than co-ordinates of a distant target expressed in degrees. To a bomber pilot, they are a point on an almost artificial world miles below. To a rocket man, they are just a release button and an order to fire.

It is only in the various kinds of infantry that closer killing occurs. An individual can be seen through the sights of a rifle, and it is difficult to close the mind to the suffering caused when using a bayonet. Any inhibitions, either from an imaginative gun-layer or a 'squeamish' infantryman, are overcome by the rigid hierarchy of the forces. One of man's greatest potential redeemers, his conscience, is satisfied with the phrase 'I was ordered to . . .'. This was possibly the most frequently heard answer to prosecution at the Nuremburg war crimes trial, the final reckoning for the leaders of the Nazi regime in 1946–47. Armed forces are great social units, lead ultimately by the leader of a country. They are full of the kind of co-operation needed by a pack to hunt, and the kind of grooming and submission needed to reinforce leadership and bonds between men. In this perspective it is hardly surprising that any inhibitions about killing are eradicated very early in a military career.

Whatever the roots of man's aggressive behaviour both in war and in peace, they present a problem to be analysed and conclusions to be applied. Probably aggression is innate in man, just as it is in many animals. Aggressive tendencies and competition between groups are not necessarily bad, because they produce strong family and social bonds. Only aggressive creatures have such bonds between individuals. But man's aggression is out of hand. The non-adaptive behaviour of war has become established in society, probably by cultural means. Man's weapons are more powerful than ever before and there are more people to be killed, should the 'ultimate threat' be carried out.

The most fatal mistake would be to write off man's nature as 'just the animal in him'; quite the reverse is true. To talk of a man behaving like an animal is to do animals a severe injustice. They have aggression under control. Furthermore, if war and other ills are blamed on inheritance and innate drives, the implication is that they cannot be cured. Such an acceptance of the situation could be dangerous.

With the help of animal behaviour studies and all they reveal of territory and aggression in the animal world, the special problems of man can be highlighted, understood and – perhaps in time – overcome.

Indexes

Numbers in italic type indicate an illustration.

Animal index

African wild dog 92
American flicker *81*
American opossum 107
American porcupine *105*
Anole 72
Ant 23, 52, 89, 107
Antelope 29, 31, 43, *44*, 46, *58*
Archerfish 86, *87*, 88
Argus pheasant *46*, 48
Armadillo *104*
Army ant 23, 89
Australian brown snake *33*
Avocet 46, *68*, 70

Baboon 14, *55*, 84
Bear *90–91*, *132*
Beetle 17, *39*, 107
Bird of paradise 48
Bird of prey 89
Bison 43, *45*, 46
Bitterling 46, 81
Bittern *66*
Blackbuck 43
Blue tit 71
Blue wren *82*
Bombardier beetle 107
Boomslang *78*
Budgerigar 81
Bullfinch 83
Bunting *44*

Camel *116*, *119*
Canary 123
Cape hunting dog 92
Caribou *48–49*
Carrier-pigeon *122*, *123*
Cat 72, *79*
Chaffinch *33*, 50
Chameleon *60*
Cheetah 89
Chevrotain 30
Cichlid fish 83
Civet 108
Cobra 18, *21*
Cockatoo *78*
Common tern *103*
Coot 35
Corn bunting *44*
Crab 17, 29, *38*, 41
Crane 69
Cricket *33*, 41
Crown of thorns starfish *94*, 95
Cuttlefish 76

Deer *2*, *16*, 29, *30*, 31, *37*, 46, *48–49*, *98–99*, 105

Dog 29, 43, *44*, *63*, 64, *65*, *90–91*, 92, *124*, *125*, 126
Domestic hen 13
Dormouse 105
Dragonfly 41
Driver ant 23
Duck 14, *22*, 48, *59*

Eagle 28, 89
Egret *80*
Eider-duck *22*
Elephant 29, *34*, 70, *120–121*, 122
Elephant seal *32*
Elk 31
Emerald lizard 81
Empid fly 96

Fence lizard 81
Fiddler crab *38*, 41
Field mouse 105
Fighting cock 17, *18*, 70
Firemouth *78*
Fish 17, *28*, 29, 35, *36*, 43, 46, *62*, 63, *69*, 70, *75*, *78*, 81, 83, 86, *87*, 88, 93, 103
Flicker *81*
Flour beetle 17
Fly 96
Fox 22, 43, 86
Frilled lizard *72*
Frog 50, *58*
Frogmouth *78*

Gannet *40*, *50*
Garden spider *96*
Gazelle 22, *44*
Giraffe 13, *15*, 33
Gnu *58*
Goanna *13*
Goat 29
Golden eagle 28
Grant's gazelle *44*
Great tit 70
Green lizard *38*
Gull *14*, *22*, 28, 29, 51, *64*, *68*, 71, *72*, 73, *74*, 81, 84, 102

Hare *31*
Harrier *103*
Hedgehog 104
Hen 13
Heron *84*
Herring gull *22*, 28, 51, *64*, *68*, 71, *72*, 73, 81, 84, 102
Hippopotamus *1*, 16
Honeybee 14, *51*, 52
Horse *19*, 29, *110–115*, *117*, *118*, 126
Howler monkey 52
Hunting wasp 102
Husky *63*, *90–91*

Indian blackbuck 43

Jack Dempsey 35
Jackal 43

Jackdaw 13, 103
Jay 22

Kagu 72
Kissing gourami 35, *36*
Kittiwake 51

Langur 13
Laughing gull *74*
Legionary ant 23
Lemming 17
Leopard *20*, *89*, 94
Lion *9*, 17, 46, *83*, 86, *88*, 89, 90
Lizard *13*, 34, *38*, *60*, *72*, 81, *104*, 105
Locust 23
Long-eared owl *67*

Mandarin duck 48, *59*
Maned wolf 72
Mangabey *73*
Mantis 14, *100*, *101*, 108
Marsh harrier *103*
Minnow 103
Mongoose 18, *21*, 108
Monkey 13, 14, 52, *55*, *73*, 84
Moose *98–99*
Mosquito 26
Mouse 105, 123
Mule 116
Muntjac 43
Musk deer *30*
Musk-ox 32, *102*, 103
Mussel 46
Mynah 35

Night heron *84*

Opossum *107*
Oryx 31
Owl 22, *67*
Oxen 116
Oystercatcher 69

Panda *45*
Pangolin *104*
Paper wasp *66*
Peacock *57*
Peccary 108
Penguin 51
Pheasant *46*, *47*, 48
Pigeon *122*
Piranha fish 93
Polar bear *90–91*, *132*
Porcupine *104*, *105*
Processionary moth caterpillar *56*
Puffin *59*

Rabbit 105
Rat 26, *33*, 71, 140, *141*, 142
Rattlesnake 33, *132*
Red deer *2*, 30, *37*, 49
Reedfrog *58*
Rhinoceros 29
Robin 75, *76*, 77
Ruff 34, *47*, 48, 72

149

Sage grouse *70*, 72
Salmon 29
Sand lizard 34
Seagull *14*
Seal *32*
Shark 93
Sheep 29, 46
Shrew 14, 84
Siamese fighting fish 17
Sika deer *16*, 30
Skunk *106*, 108
Snake 18, *21*, *33*, *132*
Spider *96*, 102
Spoonbill 72
Spotted skunk *106*, 108
Stag beetle *39*
Starfish *94*, 95, 104
Starling 51, 69
Stickleback 43, *62*, 63, *69*, 70, *75*
Striped skunk *106*, 108
Swan *54*

Termite 107
Tern 51, 69, *103*
Tiger 22, *73*, 89, *92*, 93, *97*
Tit 70, 71
Trapdoor spider 102
Tree frog 50
Tree shrew 33
Triton shellfish *94*, 95
Trumpetshell *94*, 95

Virginia opossum *107*

Warthog *9*
Wasp *66*, 102
Water deer 30
Wild cat *79*
Wolf 23, 43, *73*, 89, 91, *98–99*
Wolf spider 102
Wren *82*

Zebra *8*, 22, *58*, *88*, 103

Subject index

Aggression: 9, 28, 62, 96; man 9, *10–11*, *12*, *128–129*, 130, 139, 140, *143*, 148
Alarm calls: jackdaw 103
Anxiety posture: herring gull *68*
Appeasement gestures: baboon 84; egret *80*; herring gull 81, 84; man 139, 145; night heron *84*; shrew 84; wren 81, *82*
Attack (see Threat displays): 22, 61; methods in interspecific war 86–92
Australopithecus: *134*, *137*

Behaviour: conflict 63, 71, 73; instinctive *56*, 61; studies 54, 55, 56, 61, 63
Belling: red deer 30, *37*

'**Billing**': puffins *59*
Bird strikes on aircraft: *126*
Boredom: man 145

Camouflage: 104
Cannibals: empid fly 96; mantis 14, *101*; spider *96*
'**Charging**' in defence: 103
Chemical defence: 104; ant 107; bombardier beetle 107; civet 108; minnow 103; mongoose 108; peccary 108; skunk *106*, 108; termite 107
Chemical warfare: man *24–25*, 26
Choking: herring gull *72*, 73; laughing gull *74*
Civilisation: man 130
Communities: advantages 50; flocks 40, *50*, 51; groups 52; herds *48–49*, 50, *58*; insect colonies *51*, 52; nesting colonies 40, *50*, 51; shoals 50; societies 52
Conflict behaviour: 63, 71, 73
Courtship: 96, 140
Courtship displays: bitterling 81; blue wren *82*; bullfinch 83; cichlid fish 83; emerald lizard 81; herring gull 81; sage grouse *70*
Courtship rituals: egrets *80*; peacock *57*; puffins *59*; reedfrog *58*
Crop spraying: *24–25*
Culture: man 12, 131, 139

D.D.T.: 26
Defence methods in interspecific war: *22*, *102*, *104*, *105*; alarm calls 103; camouflage 104; 'charging' 103; chemicals 103, 104, *106*, 107, 108; flight 108; 'freezing' 105; mobbing 22, *103*; self-amputation 104, 105; shamming death *107*; threat display 108; trip wires 102; warning colours 108; weapons 104
Displacement activity: avocet *68*, 70; blue tit 70; crane 69; elephant 70; fighting cock 70; great tit 70; man 71; oyster-catcher 69; rat 71; starling 69; stickleback *69*, 70; tern 69
Displacement nest-building: herring gull 71; stickleback 70

Escape (see Threat displays): 61
Evolution: biological 11; cultural 12; sexual selection *46*, *47*, 48, *59*

Facial expressions of threat: dog 63, 64, *65*; tiger *97*
Fear (see Threat displays): 61; ruffs 35
Flight in defence: 108
Flocks: chaffinch 50; gannet *40*, *50*; starling 51

Food chains: 86
'**Freezing**': deer 105; rabbit 105
Grass-pulling: herring gull *71*
Groups: howler monkey 52

Herds: caribou *48–49*; red deer 49; zebra and gnu *58*
Hiroshima: *130*
Home (see Territory)
Hunter and hunted relationship (see Interspecific war)
Hunters: African wild dog 92; archerfish 86, *87*, 88; army ant 23, 89; birds of prey 89; Cape hunting dog 92; cheetah 89; eagle 89; leopard *89*; lion *9*, *88*, 89, *90*; tiger 89; wolf 89, 91, *98–99*
Hydrogen bomb: *136*

Insect colonies: ant 52; honeybee *51*, 52
Instinctive behaviour: processionary moth caterpillar *56*, 61
Intention movements: herring gull 64, 68; man 64
Interspecific war (between-species war): 9, 21; African wild dog 92; archerfish 86, *87*, 88; army ant 23, 89; birds of prey 89; Cape hunting dog 92; cheetah 89; cobra and mongoose *21*; crown of thorns starfish *94*, 95; eagle 89; eider-duck and herring gull *22*; fox 86; husky and polar bear *90–91*; lion 89, 90; lion and warthog *9*; lion and zebra 22, *88*; man and animals *24–25*, 26; mantis and moth *100*; piranha fish 93; tiger 89; tiger and gazelle 22; triton shellfish *94*; wolf 89, 91; wolf and moose *98–99*
Intraspecific war (within-species war): 46; Australian brown snakes *33*; baboons 14; bees *51*; chaffinches *33*; coot 35; crickets *33*, 41; dragonfly 41; ducks 14; elephants *34*; elephant seals *32*; fiddler crabs 41; fighting cocks *18*; fish 28; giraffes *13*, *15*, 33; goannas *13*; green lizards *38*; gulls 14; hares *31*; horses *19*; Jack Dempsey 35; kissing gouramis 35, *36*; lion *83*; musk-ox 32; mynah 35; oryx 31; polar bears *132*; rats *33*, *141*; rattlesnakes 33, *132*; red deer *2*, 30, *50*; rhinoceros *29*; ruffs 34, *47*; sand lizards *34*; seagulls *14*; shrews 14; sika deer *16*; stag beetles *39*; tree shrews 33; zebras *8*

Killing: hippopotamuses 16; honeybees 14; mantis 14

Lorenz, Professor Konrad: 54, 74

Man: 8, 9, *10–11*, *12*, 25, 26, 64, 71, 86, 93, 94, 95, *128–129*, *143*, *144*, *146–147*; aggression 9, *10–11*, *12*, *128–129*, 130, 139, 140, *143*, 148; appeasement gestures 139, 145; boredom 145; chemical warfare *24–25*, 26; civilisation 130; culture 12, 131, 139; destruction *130*; inhibitions 137, 138; mechanical warfare 26; overcrowding 140, *142*; rituals 134, 139, 145; societies 52; submissive gestures 139, 145; suicide 140, *142*; territory 28, *138*, 139; threat displays *135*, *136*, *145*; weapons *133*, 134, *136*, 138, 145
'Man-ape': 134, *137*
Man-eaters: leopard 94; piranha fish 93; shark 93; tiger *92*, 93, *97*
Mechanical warfare: man 26
Mobbing: jackdaw 22, 103; jay 22; terns *103*; zebra 103

Nesting colonies: gannets *40*, *50*; gull 51; kittiwake 51; penguin 51; tern 51

Overcrowding: crab 17; fish 17; flour beetles 17; hippopotamuses 16; lemming 17; lion 17; man 140, *142*; rats 140, *141*; sika deer *16*

Peck-order: 84; domestic hens 13; jackdaws 13; langurs 13; monkeys 13
Pedal gland: muntjac 43, *44*
Pests: mosquito 26; rat 26
Phalanx: musk-oxen *102*
Physiological changes in threat display: *70*, 71, *72*
Pollution: 26
Predator and prey relationship (see Interspecific war)
Preening movements: crane 73
Preparatory movements: herring gull 64, 68; man 64
Psychological territory landmarks: 46; stickleback *62*, 63

Redirected movement: herring gull *71*
Releasers (signals): American flicker *81*; budgerigar 81; cuttlefish 76; fence lizard 81; mangabey *73*; night heron *84*; robin 75, *76*, *77*; stickleback 74, *75*; tiger *73*; wolf 73
Reproductive fighting (see Intraspecific war)
Ritualisation of threat displays: 73, *74*
Rituals: 134; courtship *57*, *58*, *59*, *80*; man 134, 139, 145

Rival fighting (see Intraspecific war)
Rules: 33
Rutting battles: red deer *2*, 30, 50

Sand-digging: stickleback 70
Self-amputation: dormouse 105; field mouse 105; lizard 104, 105; starfish 105
Sexual selection: Argus pheasant *46*, 48; bird of paradise 48; mandarin duck 48, *59*; pheasants 47; ruff *47*, 48; sheep 46
Shamming death: American opossum 107; Virginia opossum *107*
Shoals: 50
Signals and signalling devices (see Releasers)
Skinner, Dr.: 54
Society: man 52
Staged fights: cobra and mongoose 18, *21*; fighting cocks 17, *18*; Siamese fighting fish 17, 18
Stress: 17
Submissive gestures: baboon 84; egret *80*; herring gull 81, 84; man 139, 145; night heron *84*; shrew 84; wren 81, *82*
Suicide: man 140, *142*
Survival in extreme environments: 25

Territorial disputes (see Intraspecific war): 13, 28
Territory: 41, 42, 43, 46, 51, 61; gannets *40*, *50*; golden eagle 28; herring gull 28; kob *42*; man 28, *138*, 139; stickleback *62*
Territory landmarks: psychological 46, *62*, 63; scent 43, *44*, *45*; signs *138*; vocal 43, *44*, 52
Territory marking: Alsatian dog *44*; bison 43, *45*; corn bunting *44*; dog 43, *44*; fox 43; Grant's gazelle *44*; Indian blackbuck 43; jackal 43; man *138*; muntjac 43, *44*; panda *45*; red deer *37*; wolf 43
Threat displays (animals): 12, 13, 54, 61, 63; baboon *55*; bittern *66*; boomslang *78*; chameleon *60*; cockatoo *78*; fiddler crabs *38*; firemouth *78*; frilled lizard *72*; frogmouth *78*; hippopotamus *1*; long-eared owl *67*; man *135*, *136*, *145*; mantis 108; paper wasp *66*; robin *76*, *77*; sage grouse *70*; swan *54*; wild cat *79*
Threat displays (within the territory): anxiety posture *68*; choking *72*, 73, *74*; displacement activity *68*, *69*, 70, 71; facial expressions *63*, 64, *65*; grass-

pulling *71*; in courtship 81, 83; intention movements 64, 68; physiological changes *70*, 71, *72*; preening movements 73; preparatory movements 64, 68; redirected movement *71*; releasers *73*, 74, *75*, *76*, *77*, *81*; ritualised 73, *74*; sand-digging 70; upright threat posture *64*, 68
Threat signals (see Releasers)
Tinbergen, Professor Niko: 64, 74
Trip wires used as defence: trapdoor spider 102

Upright threat posture: herring gull *64*, 68

War animals: camel *116*, *119*; canary 123; carrier-pigeon *122*, 123; dog *124*, *125*, 126; elephant *120–121*, 122; horse *110–115*, *117*, *118*; mouse 123; mule 116; oxen 116
Warfarin resistance: rats 26
Warning colours: skunk *106*, 108
Weapons: 33, *133*, 134, *136*, 138, 145; antlers *2*, 29, 30, 46, 104; armour *104*; chemicals 103, 104, *106*, 107, 108; claws 29, 104; hoofs 29; horns 29, 30, 46, 104; quills 104, *105*; teeth 29, 104; tusks 30

We would like to acknowledge permission to reproduce the following: an extract from 'Skunks' Copyright 1954 by Robinson Jeffers. Reprinted from Selected Poems, by Robinson Jeffers, by permission of Random House Inc; an extract from 'A letter to a few good poets I know' by Dave Cunliffe, copyright BB Bks of Blackburn.

Photo Credits

Cover – Albert Visage: Jacana
1 – Ivor Jarman
2 – Roy A Harris & KR Duff
7 – WT Miller
8 – Okapia
9 – WT Miller
10,11,12 – Popperfoto
14 – Barnabys
15 – RS Virdee
16,18 – Okapia
19 – Margaret Wünsch
20 – Bavaria
21 – Okapia
24,25 – Keystone
27 – WJC Murray: NHPA
29 – Toni Angermayer
30 – Popperfoto
31 – Roy A Harris & KR Duff
32 – John Warham
34 – Arthur Christiansen
36 – Jane Burton: B Coleman
37 – roebild
38T – Jane Burton: B Coleman
   B – Peter Hill
39 – Bavaria
40 – Eric Hosking
44T – Peter Hill
   BL – roebild
   BR – SC Porter: B Coleman
45 – Zool Soc London

47 – Arthur Christiansen
48,49,50 – Fred Bruemmer
51 – Stephen Dalton: NHPA
53 – Wolfgang Lummer
54 – roebild
55 – Toni Angermayer:
   B Coleman
56 – Graham Pizzey: B Coleman
57 – Animal Photography
58T – Root/Okapia
   B – Jane Burton: B Coleman
59T – G Rüppell
   B – Zool Soc London
60 – Heather Angel
63 – Fred Bruemmer
66T – Anthony Bannister: NHPA
   B – Douglass Baglin: NHPA
67 – Jane Burton: B Coleman
70 – Joe Van Wormer:
   B Coleman
72 – Popperfoto
76 – Peter Hinchliffe: B Coleman
77 – H Schrempp
78T – LE Perkins
   L – John Markham
   CR – Joe B Blossom: NHPA
   BR – M Boorer
79 – G Kinns: AFA
80 – KB Newman
82 – John Warham
83 – RS Virdee
85 – Zool Soc London
88 – Popperfoto
89 – CAW Guggisberg:
   B Coleman

90,91 – Fred Bruemmer
92 – Mary Evans Picture Library
94 – Ben Cropp
96 – Stephen Dalton: NHPA
97 – roebild
98,99 – L David Mech
100 – Anthony Bannister: NHPA
101 – PH Ward
102 – Fred Bruemmer
103 – Eric Hosking
109 – By Courtesy of the
   Trustees of the British Museum
114,115 – Ullstein
116 – Imperial War Museum
117 – Musée National D'Art
   Moderne
119 – Imperial War Museum
120,121 – Mansell Collection
122,124 – Imperial War Museum
127 – Popperfoto
128,129 – Camera Press
130 – Keystone
132T – Jeffrey C Stoll: Jacana
   B – San Diego Zoo
133 – Barnabys
135 – Transworld
136 – Popperfoto
137 – British Museum (Natural
   History)
138 – Keystone
141 – Okapia
142 – Cathy Jarman
143 – Keystone
144,145 – Camera Press
146,147 – Barnabys